The dynamic liquid state

for June, Rowena and Charmian

The dynamic liquid state

A. F. M. Barton

Senior Lecturer in Physical Chemistry
Murdoch University, Western Australia

Longman

Longman
1724-1974

Longman Group Limited
London
Distributed in the United States of America by
Longman Inc., New York
Associated companies, branches and representatives
throughout the world

© Longman Group Limited 1974

First published 1974

ISBN 0 582 44276 1
Library of Congress Catalog Card Number: 73-90500

Filmset by Keyspools Ltd, Golborne, Lancs
Printed in Great Britain by
Whitstable Litho Ltd., Whitstable

Contents

Preface

This book is intended as a general introduction to the subject of the liquid state, influenced by my belief in the importance of understanding the dynamic nature of liquid behaviour. In order to limit the book's length I have concentrated on the properties of *pure* liquids, and omitted discussion of liquid mixtures and solutions. Topics such as glasses and the melting process have also been excluded.

It has been my intention to present the subject at a level which is simple but at the same time adequate to describe current concepts and the results of recent research and to give an *appreciation* of the liquid state, in the same way that an appreciation of music does not require an ability to play a musical instrument.* Additional material not essential for the understanding of later chapters is included in smaller type, and the annotated bibliographies should allow easy access to the recent literature. In particular, Chapter 12 on "Theories of the Liquid State" is a very brief account of a subject which continues to receive considerable attention, but in compensation for this brevity there is an extensive bibliography.

I would like to acknowledge gratefully the contributions of many authors. Instead of referring to each specifically, I have collected together at the end of each Chapter "general references" which provide sources of more detailed information and discussion. A "star" code in the general references indicates the order in which the accounts should be read, from one star for a "popular" account to three stars for an advanced or specialized treatment. When references are made in the text to particular research papers, specialized techniques and the origins of important concepts, these may be found in the "specific references" also found at the end of chapters.

A considerable proportion of this book was written while I was on leave from Victoria University of Wellington. I would like to express my appreciation for this period of leave, and also for the hospitality extended by the Chemistry Department, Imperial College of Science and Technology, University of London, and the assistance I received from the British Council. I wish to thank also my colleagues at Victoria University for their critical reading of the manuscript, Mrs C. Mitchell for her patient typing, and Mr E. F. Stevens for careful preparation of the drawings.

* Kohn, H. W. (1972), *J. Chem. Educ.*, 49, 728.

Chapter 1
Introduction to the liquid state

The "structure" of a liquid on a molecular scale is considerably more difficult to visualize than the complete disorder or randomness which occurs in a gas, or the regular arrangement which is found in a crystal. In gases, at least at the ideal limit under conditions of low density, the molecular mean free path (i.e. the distance travelled between collisions) is long compared with the molecular diameter. Disorder due to thermal motion predominates over the ordering imposed by intermolecular forces. Under these low density conditions the molecules interact mainly by two-body collisions, and simultaneous collisions involving three or more molecules are rare. On the other hand, in an ideal crystal ordering predominates and the vibrational amplitude is small compared with the diameter. Any one molecule is always interacting with many neighbours. Both these limiting cases are relatively simple to treat mathematically. Liquids, however, maintain a structure which is a compromise between order and disorder and for this reason, despite a tremendous effort devoted to the analysis of liquid structure by both experimental and theoretical methods, neither physical "pictures" of the arrangement of a molecule in a liquid nor quantitative theories explaining liquid properties are as far advanced as those of the solid and gaseous states.

The *condensed* phases, liquids and solids, resemble each other in the property of *cohesion*; they both maintain a boundary surface, but in liquids the cohesive forces are not strong enough to prevent translational movements of individual molecules. Liquids and solids also have a similar degree of close packing: their densities differ in general by less than 15%, and in fact in a few cases such as the metal bismuth, the salt rubidium nitrate, and water, the rather "open" solid structure becomes more close-packed on melting.

There are two fundamental problems in discussing the structure of liquids. The first is that of understanding the nature of molecular interaction. (The word "molecule" is used for simplicity to indicate "atom, molecule, or ion".) In general terms it can be said that the forces are repulsive for small molecular separations and that they become attractive at larger separations, but a detailed knowledge of the nature of these forces is necessary for the understanding of liquid properties. The second fundamental problem is that of relating the bulk or macroscopic properties of a system to the microscopic or molecular properties and in particular to the potential energy function which describes the way in which an isolated pair of molecules interacts. The difficulties faced in discussing liquid structure are the same as those in solvation phenomena which were emphasized recently by Friedman (1973): because the chemical bond is the central concept in chemistry, our approach to it has become relatively sophisticated. In contrast, our knowledge of *intermolecular* forces is poorly developed.

The extent to which the intermolecular potential energy controls the movement of molecules depends on the strength of movement (i.e. the kinetic energy). At the one extreme of low temperatures the mean potential energy is greater than the mean kinetic energy, and the material is an ordered solid. The other extreme, shown by dilute gases, is characterized by a mean kinetic energy much greater than the mean interaction potential energy, and complete disorder. In a liquid the molecules are sufficiently close together for there to be local ordering and only a small degree of compressibility, but the kinetic energy is large enough to prevent long-range ordering.

In considering the degree of order in a liquid, it is valuable to consider the results of X-ray diffraction experiments. The X-ray diffraction pattern of a solid *single crystal* consists of symmetrically placed points of high intensity. In the X-ray diffraction patterns of crystalline *powders*, it is found that as the particle size of the powder decreases the peaks increase in width and become diffuse, until at very small particle size the diffraction maxima become completely blurred out. If a liquid had no regularity of structure, it should give a continuous scattering of X-rays without maxima. In fact, there are a few maxima and minima, and these occur at distances corresponding to those in some ordered close-packed structures. This is interpreted as showing that liquids possess a certain amount of *short-range order*, but *long-range disorder*. The existence of some similarity in the X-ray diffraction pattern in liquids and solids should not be considered as evidence that a liquid is more like a solid than a gas. Short-range order is a necessary consequence of the similar high packing densities. Gases at low density have no X-ray diffraction pattern (except that due to internal molecular structure) but if they are compressed to high density, then a pattern typical of a liquid appears.

Because the structure of the liquid state has proved to be very difficult to describe, either qualitatively or quantitatively, several methods have been developed with very different points of view, and these approaches are summarized here. They will be discussed in greater detail later, when their significance should become more apparent. At this stage it is necessary to clarify one point: the way in which the terms *time-independent* or *static* and *time-dependent* or *dynamic* will be used. The former description refers not only to the thermodynamic or equilibrium properties, but also to the steady-state transport coefficients such as the self-diffusion coefficient and shear viscosity coefficient at zero shear rate which in many ways resemble intensive thermodynamic variables of state. They are measured over times and distances which are long on a molecular scale, and their values do not change with time. The latter description, dynamic, refers to properties whose measured values depend on the time scale or frequency of observation, and describe, for example, how, on average, the environment of a typical molecule changes with time.

The time scale as well as the distance scale is of fundamental importance when discussing liquids. It is often said that the liquid state has properties "intermediate" between those of a solid and of a gas. This is most nearly true if time is included in the consideration of properties. Solids have a structure which as well as being ordered is time-independent. At short times a liquid behaves as a solid with elastic properties, and at short distances a liquid has some degree of order. A gas on the other hand has no structure or order no matter how short the time or distance scale.

To oversimplify,

solids are always "solid"

gases are always "fluid"

liquids are "solid-like" at short times and short distances but "fluid-like" at long times and long distances.

All the properties of the liquid state determined experimentally are bulk properties, i.e. they represent a combination or averaging of molecular properties. The link between measured or macroscopic and molecular or microscopic properties is statistical thermodynamics, the science dealing with large numbers of particles. This is discussed in Chapter 12, "Theories of the liquid state", but the greater part of this book is concerned with the way in which experimentally accessible properties have been described and interpreted to yield parameters which permit an appreciation of the fundamental properties of liquids.

3

Static properties

(i) Thermodynamics

Equilibrium properties such as volume are measured as a function of pressure and temperature, described by equations of state, and used to evaluate other thermodynamic parameters such as the internal pressure which may reflect more clearly what is happening on a molecular scale.

(ii) Fluid mechanics

The transport coefficients may be described with the assumption of a fluid continuum, i.e. the molecular properties are neglected. If a disturbance is varying slowly in space and time, the response may be described in terms of thermodynamic quantities and transport coefficients. For example, if the local density in a liquid is disturbed by the passage of a sound wave, the behaviour is then described in terms of the thermal conductivity, the shear viscosity and the bulk viscosity. The period of a sound wave is very much longer than the time scale of a molecular structure rearrangement, and the sound wavelength is very much longer than an intermolecular distance. This is the hydrodynamic limit.

(iii) Gas-like properties

These considerations are based on the "continuity" of the fluid phases, i.e. by adjusting the pressure and temperature to super-critical conditions, a liquid and gas may be interconverted without passing through a phase transition.

(iv) Molecular distribution functions

Results of some types of radiation scattering experiments, for example X-ray diffraction, provide time-averaged information on molecular positions, and this is described in terms of a distribution function.

(v) Electromagnetic properties

These properties are important in liquids with polar or polarizable molecules.

Dynamic properties

(i) Spectroscopic techniques, yielding time-dependent distribution functions

Most spectroscopic techniques (both absorption and scattering) provide "relaxational" or time-dependent information on averaged relative molecular positions, and molecular translational and rotational diffusion. Although this information is on a molecular distance scale, it should be noted that the experimental probes (the beams of radiation or particles) do not provide a picture of the behaviour of one single molecule, but an average over a very large number of molecules. Despite this, the fact that the time scale is of the order of magnitude of molecular reorientation and diffusion times means that this information is potentially more valuable for determination of the nature of molecular interactions than results of static measurements. If one considers this from the point of view of the frequency of the spectroscopic radiation, the static measurement corresponds to one point at zero frequency while the dynamic method provides a spectrum over a frequency range.

(ii) Solid-like properties

At high frequencies of mechanical perturbation (hypersonic frequencies) there is evidence of elastic behaviour. In addition there is evidence from radiation scattering of the existence of phonons.

In the following chapters the thermodynamic, fluid mechanic, distribution function, electromagnetic and spectroscopic approaches to the liquid state will be introduced in turn with emphasis on the way in which each reflects one aspect of the intermolecular properties of liquids. To obtain the most complete picture of the liquid state it is necessary to combine all of these various points of view and to avoid the common mistake of becoming preoccupied with one of them.

General references

* BARTON, A. F. M. (1973), "The Description of Dynamic Liquid Structure", *J. Chem. Educ.*, 50, 119–22.

** CHEN, S-H. (1971), "Structure of Liquids", in Henderson (1971), Chap. 2, pp. 85–156.

COVINGTON, A. K. and DICKINSON, T. eds. (1973), *Physical Chemistry of Organic Solvents*, Plenum, N.Y.

★ DREISBACH, D. (1966), *Liquids and Solutions,* Houghton Mifflin, Boston, U.S.A. Part 1.
One of a "classic researches" series, presenting developments largely by means of extracts from original papers.

★★★ EGELSTAFF, P. A. (1967), *An Introduction to the Liquid State,* Academic Press, London.

★ EYRING, H., HILDEBRAND, J. H. and RICE, S. (1963), "The Liquid State", *International Science and Technology,* No. 15, 56–66.

★★ FLOWERS, B. H. and MENDOZA, E. (1970), *Properties of Matter,* Wiley, London, particularly Chap. 7, pp. 176–224.

★ FRIEDMAN, H. L. (1973), "Modern Advances in Solvation Theory", *Chem. in Britain,* 9, 300–5.

★★★ FRISCH, H. L. and SALSBURG, Z. W. (1968), eds., *Simple Dense Fluids,* Academic Press, New York.
A collection of specialized chapters by different authors, with discussion and data on thermodynamic and transport properties, radiation scattering, dielectric data and nuclear and electronic spectroscopy of the simpler liquids.

★★ HENDERSON, D. (1971) ed., *Liquid State,* Vol. 8A in *Physical Chemistry: An Advanced Treatise,* eds. Eyring, H., Henderson, D. and Jost, W., Academic Press, New York.
A collection of chapters at a moderately advanced level with a variety of points of view.

★★★ HIRSCHFELDER, J. O., CURTISS, C. F. and BIRD, R. B. (1954), *Molecular Theory of Gases and Liquids,* Wiley, New York.
A standard work on gases and liquids from a molecular and statistical thermodynamic viewpoint.

★★ HUGHEL, T. J. (1965) ed., *Liquids: Structure, Properties, Solid Interactions,* Proc. Symposium, Warren, Michigan, 1963, Elsevier, Amsterdam.

KOHLER, F. (1972), *The Liquid State,* Vol. 1 in *Monographs of Modern Chemistry,* Verlag Chemie, Weinheim.

★★ MOELWYN-HUGHES, E. A. (1957), *Physical Chemistry,* Pergamon, London, Chap. 16, pp. 675–735.

★ MOORE, W. J. (1972), *Physical Chemistry,* Longman, London, 5th edn., Chap. 19, pp. 902–27.

*** MOUNTAIN, R. D. (1970), "Liquids: Dynamics of Liquid Structure", *Critical Rev. Solid State Sciences,* 1, 5–46.
A valuable review at an advanced level.

** PRYDE, J. A. (1966), *The Liquid State,* Hutchinson, London.
An introductory treatment from the point of view of "static" rather than "dynamic" properties.

** ROWLINSON, J. S. (1969), *Liquids and Liquid Mixtures,* Butterworths, London, 2nd edn.
A useful account of the thermodynamic and mixing properties of liquids.

** ROWLINSON, J. S. (1970a), "Structure and Properties of Simple Liquids and Solutions: A review", *Disc. Faraday Soc.,* 49, 30–42.

** ROWLINSON, J. S. (1970b), "The Structure of Liquids" in *Essays in Chemistry,* eds. Bradley, J. N., Gillard, R. D. and Hudson, R. F., Academic Press, London, Vol. 1, pp. 1–24.
A very helpful introduction to the subject.

** SCOTT, R. L. (1971), "Introduction" in Henderson (1971), Chap. 1, pp. 1–84.
A useful discussion at a reasonable level of many aspects of the description of the liquid state.

* SMITH, B. L. (1971), *The Inert Gases, Model Systems for Science,* Wykeham, London, Chap. 6, "Liquids".

*** TEMPERLEY, H. N. V., ROWLINSON, J. S. and RUSHBROOKE, G. S. (1968), eds., *Physics of Simple Liquids,* North-Holland, Amsterdam.
A comprehensive collection of chapters at a rather advanced level.

* WOODHEAD-GALLOWAY, J. (1972), "Towards the Structure of Liquids", *New Scientist,* 56, 399–403.
A general account at a popular level.

Chapter 2
Thermodynamics and equations of state

The thermodynamics of matter in equilibrium describes a state, which is supposed to persist indefinitely unless disturbed, in terms of a number of variables such as pressure p, molar volume V, temperature T, molar internal energy U and molar entropy S, which are discussed in textbooks of thermodynamics and general physical chemistry. The relation linking a set of variables for a particular liquid is an "equation of state" for that liquid.

Virial equations of state

Equations of state which involve only pressure, volume and temperature may be described as "virial" equations of state, using the word "virial" in its general sense:

$$f(p, V, T) = 0 \qquad (2.1)$$

where $f(p, V, T)$ is a function to be determined experimentally. Detailed descriptions of equations of state have been given by Macdonald (1966, 1969, 1971). For an ideal gas, the virial equation of state is

$$\frac{pV}{LkT} - 1 = 0 \qquad (2.2)$$

where L is the Avogadro constant. For a real gas it is customary to use the expression

$$\frac{pV}{LkT} - 1 = F(T, V) \qquad (2.3)$$

where $F(T, V)$ can be written as a power series expansion in reciprocal molar volume, involving the virial coefficients B_j:

$$F(T, V) = \sum_{j \geqslant 1} B_{j+1} V^{-j} \qquad (2.4)$$

This is the equation usually called "the virial equation". At limiting low densities in a gas, molecules can be considered to interact only in pairs, and the probability of a pair of molecules occurring at a given separation r is determined only by the Boltzmann factor of their mutual energy. At high densities, however, this probability is influenced also by the presence of other molecules with which these two molecules can interact. Consequently the expansion of the pressure (and also molecular correlation functions, as discussed in Chapter 12) in terms of powers of the number density ρ has played an important part in the development of the theory of fluids. The coefficients are the second, third, fourth, etc., virial coefficients, and formally depend respectively on the interactions of two, three, four, etc., molecules considered as isolated groups. The first few terms of the expansion of p provide a reasonable equation of state for gases at low density (below the critical density), but the convergence of the expansion is slow. For high density fluids the expansion diverges and is of little practical use. Equation (2.2) may be written also in the form

$$ p = \frac{LkT}{V} \tag{2.5} $$

and should be compared with the van der Waals equation of state (van der Waals, 1873),

$$ [p + a(L/V)^2][V - Lb] = LkT = RT \tag{2.6} $$

or

$$ p = \frac{LkT}{V - Lb} - a\left(\frac{L}{V}\right)^2 \tag{2.7} $$

Both are special cases of the general class of equations of state,

$$ p = \frac{LkT}{V}B\left(\frac{L}{V}\right) - A\left(\frac{L}{V}\right) \tag{2.8} $$

where B and A are functions of the number density L/V but are independent of temperature and pressure. This equation has been called the generalized van der Waals equation (Rigby, 1970). The van der Waals equation is now 100 years old, but it introduced a concept which is still the basis of most approaches to the liquid state: the separation of the tendency of molecules to (i) attract each other at moderate distances and (ii) repel each other at short distances. It is now known (Chapter 12) that the *solid–fluid* phase transition would occur even in the absence of attractive forces (providing there was some external device to hold the material together), and the detailed structure of the liquid state is strongly

9

dependent on the repulsive forces. The occurrence of a *liquid–gas* phase transition does depend on the existence of attractive forces, but attractive forces tend to be less specific in their effects. The resultant of these forces may be considered as a large negative potential energy responsible for the existence of the liquid state, or as an "internal pressure" holding the liquid together.

Equation (2.6) was designed to correct the perfect gas equation (2.2) for the two factors of attraction and repulsion.

(i) The "a" term was intended to correct for the attractive forces between molecules. Van der Waals assumed that the net force from all neighbours on a molecule in the body of a dense fluid is zero. The potential energy is lowered by the interaction, but is uniform so there are no resultant forces. (The nature of this potential energy may be described mathematically: Rowlinson, 1970.) This correction corresponds to an attractive force long range compared with molecular diameters, although in fact intermolecular forces fall off much more rapidly with distance.

(ii) The "b" term was introduced to allow for the finite size of the molecules, each being considered to have a hard core (the collision diameter). This procedure may now be interpreted as an attempt to allow for the repulsive forces in the interaction.

It should be noted that in the van der Waals equation, assumed *molecular* properties are converted into predicted *bulk* properties by the simple expedient of multiplying by the number of molecules (look at the $(V - Lb)$ term in equation (2.7)). This is, of course, the very simplest method of making the transfer from molecular to bulk properties, and it cannot be expected to work very well. Despite this simplicity of approach, the van der Waals equation does describe in approximate terms the properties of liquids and compressed gases. After a period of neglect, there has been a revival of interest in the van der Waals equation to determine why it should work better than might be expected.

The corrections to this equation of state required for real liquids are small, and this has been shown to be due to the insensitivity of the heat capacity to volume over the entire liquid range at all temperatures, even for short range forces (Alder, 1972). However, this heat capacity correction cannot be neglected for the calculation of internal energy and other thermodynamic functions.

Thermodynamic equations of state

Among the important thermodynamic equations for systems of constant composition and total mass are the "thermodynamic equations of state":

$$(\partial U/\partial V)_T = -p + T(\partial p/\partial T)_V \qquad (2.9)$$

$$(\partial H/\partial p)_T = V - T(\partial V/\partial T)_p \qquad (2.10)$$

These equations relate the internal energy (U) and enthalpy (H) changes with volume or pressure to the p–V–T behaviour and are therefore equations of state which have a thermodynamic basis.

Equation (2.9) may be expressed

$$p = T(\partial p/\partial T)_V - (\partial U/\partial V)_T \qquad (2.11)$$

and comparison with equation (2.8) shows that liquids which obey the generalized van der Waals equation have values of $(\partial p/\partial T)_V$ and $(\partial U/\partial V)_T$ which are functions only of the molar volumes, and at constant volume the pressure should increase linearly with temperature. Because $(\partial p/\partial T)_V$ and $(\partial U/\partial V)_T$ follow this simple behaviour in many cases, the quantities have been given special names and symbols:

internal pressure: $\pi = (\partial U/\partial V)_T$

isochoric (constant volume) thermal pressure coefficient:

$$\beta = (\partial p/\partial T)_V$$

so

$$\pi = T\beta - p \qquad (2.12)$$

Not only are π and β often simple functions of the volume only, but they can be determined directly by experiment.

The chorostat

The devices used to determine isochoric thermal pressure coefficients and thence the internal pressure have been called "constant volume piezometers" or "constant volume dilatometers" depending on whether they were considered constant temperature or constant pressure devices, but they are in fact "chorostats": devices to maintain a constant volume, in the same way that thermostats maintain a constant temperature and barostats maintain a constant pressure. The example shown in Fig. 2.1 was designed for use with electrically conducting liquids, molten alkali nitrates, which have melting points between 300°C and 400°C (Barton *et al.*, 1970). The Pyrex chorostat, when filled with the liquid, was inserted in a gas-pressurized pressure vessel, which was itself inside a thermostat. The liquid level was determined by an electrical indicator circuit consisting of platinum wire contacts sealed into capillary tubing and connected externally by relatively high resistances. When the liquid level rose due to increasing temperature or decreasing pressure, successive

Fig. 2.1 Chorostat for the determination of the p–V–T properties of molten alkali nitrates (after Barton *et al.*, 1970).

probes were shorted by the low resistance melt, and this sudden decrease in electrical resistance was recorded. When corrections were made for the relatively small variations with pressure and temperature of the container dimensions, a series of pairs of pressures and temperatures were determined which defined a certain liquid specific volume or density. The long, coiled capillary "neck" was included to minimize the diffusion of the pressurizing gas into the liquid.

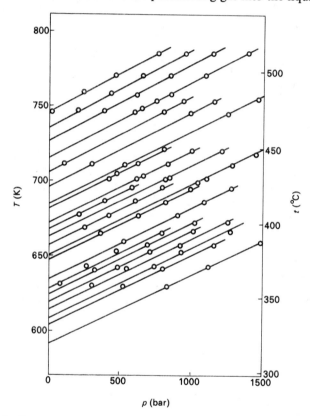

Fig. 2.2 Temperature–pressure isochores for molten sodium nitrate (after Barton *et al.*, 1970) (1 bar = 10^5 Pa).

The resulting temperature–pressure isochores for molten sodium nitrate are shown in Fig. 2.2. Such isochores are often linear within experimental error for a variety of types of liquid (Rowlinson, 1969), i.e. they obey the generalized van der Waals equation (2.8). From the slope of these graphs at any pair of (p, T) values, the internal pressure π may be evaluated from equation (2.12). A typical plot of

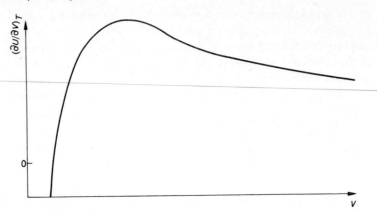

Fig. 2.3 The variation of internal pressure with volume for a liquid obeying the generalized van der Waals equation (2.8).

internal pressure against volume for a liquid obeying equation (2.8) is shown in Fig. 2.3.

It is interesting to note that if the van der Waals attraction term, $a(L/V)^2$, is identified with the internal pressure,

$$\pi = a(L/V)^2 \qquad (2.13)$$

it follows that

$$1/\beta = (V - Lb)/(Lk) = V/(kL) - b/k \qquad (2.14)$$

i.e. $1/\beta$ should be a *linear* function of the volume. This is observed in several simple liquids (Haward, 1966).

The significance of internal pressure in liquids

There are several advantages in considering the p–V–T properties of liquids in terms of the internal pressure:

(i) the internal pressure is defined thermodynamically by equation (2.9), not empirically as in other equations of state;

(ii) the internal pressure may be measured directly by the use of equation (2.12) and a chorostat;

(iii) for many simpler liquids the internal pressure depends only on the molar volume, i.e. a plot of internal pressure against volume is independent of pressure and temperature.

(iv) the balance between repulsive and attractive interactions is apparent from graphs of internal pressure against volume, and a direct, visual comparison may be made between different liquids.

(v) the internal pressure of a liquid is closely related to its *solubility parameter*, which describes the way in which it interacts with other liquids (see below).

This approach was described by Hildebrand and Scott (1950, 1962) and Hildebrand *et al.* (1970) for mixtures of liquids, although its application to pure liquids has been limited (Barton, 1971). The internal pressure is the cohesive force which is the resultant of forces of attraction and forces of repulsion between molecules in a liquid. Although strictly speaking internal pressure is a concept which can be used with precision only for a generalized van der Waals liquid (a liquid obeying equation (2.8)) when the internal pressure at a particular volume is independent of temperature and pressure, many liquids do approximate to this behaviour and comparison of the way they deviate from it is also useful.

The very simple assumptions will now be made that the molar internal energy of a liquid depends only on the volume and that it may be expressed as the sum of a negative term (favouring stability of the liquid state relative to the gas state, associated with attractive forces) and a positive term (opposing stability of the liquid state, associated with repulsive forces):

molar energy:

$$U = -\frac{A}{V^\alpha} + \frac{B}{V^\beta} \tag{2.15}$$

internal pressure:

$$(\partial U/\partial V)_T = \frac{A'}{V^{\alpha+1}} - \frac{B'}{V^{\beta+1}} \tag{2.16}$$

A, B, A', B', α and β are all positive. Figure 2.4 illustrates these expressions for molar internal energy and internal pressure, with the parameters chosen to fit the p–V–T data for ether from a liquid volume of 73.4 cm^3 mol^{-1} at 12 kbar (1 kbar = 10^8 Pa) to a volume of 400 cm^3 mol^{-1} in the vapour (Hildebrand, 1929). (There are, of course, superior empirical equations – with more parameters – for fitting ether p–V–T data, but equation (2.16) fits the data reasonably well.) The representation of the internal pressure by the algebraic sum of positive and negative terms in this way holds quite well, particularly in dilute gases and highly compressed liquids. It can be

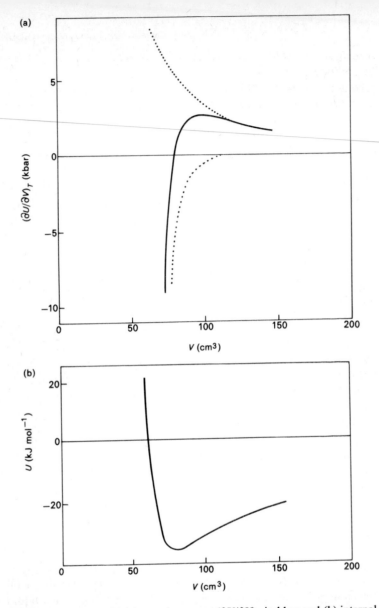

Fig. 2.4 Variation of (a) internal pressure $(\partial U/\partial V)_T$ in kbar and (b) internal energy U in kJ mol^{-1} with volume V in cm^3 mol^{-1}. The functions (equations (2.15) and (2.16)) are

$$(\partial U/\partial V)_T = 3.22 \times 10^4/V^2 - 4.88 \times 10^{19}/V^{10}$$

$$U = -3.22 \times 10^4/V + 5.43 \times 10^{17}/V^9$$

in which the parameters are chosen to fit the $(\partial U/\partial V)_T$ against V data for ether. The dotted curves in (a) show the cohesion and repulsion contributions to the internal pressure (after Barton, 1971).

seen from the plot of internal pressure against volume (Fig. 2.4a) that in the region of the maximum and at higher volumes the internal pressure may be approximately represented by the attractive term only (denoted by a subscript "a").

$$U = U_a = -AV^{-\alpha} \tag{2.17}$$

$$\pi = \pi_a = \alpha A V^{-\alpha-1} = \alpha U_a V^{-1} \tag{2.18}$$

The "attractive" part of the internal energy is equal to the molar energy of vaporization, ΔU_{vap}, for vaporization at low gas pressure, so

$$\pi_a = \alpha(\Delta U_{vap}/V) = \alpha\delta^2 \tag{2.19}$$

where $\delta = (\Delta U_{vap}/V)^{1/2}$ is the *solubility parameter*, and for liquids which follow the van der Waals equation $\alpha = 1$, i.e. $\pi \propto V^{-2}$ (see equation (2.13)), so

$$\pi = (\partial U/\partial V)_T \simeq (\Delta U_{vap}/V) \tag{2.20}$$

The quantity $(\Delta U_{vap}/V)$ is known as the cohesive energy density, and is a measure of the total molecular cohesion per unit volume, while the internal pressure is the instantaneous, isothermal volume derivative of the *total* internal energy, both cohesive and repulsive. Thus the internal pressure provides one way of estimating the solubility parameter, but the use of equation (2.19) implies the unrealistic assumption that equation (2.18) holds over the entire range of liquid densities, and is unsatisfactory for more complex molecules and for higher densities.

As well as considering such empirical equations for expressing internal pressures, it is informative to study experimental pressure against volume graphs directly and to discuss their characteristics qualitatively. The plots for a variety of liquids are shown in Fig. 2.5, and further information is provided in Table 2.1. Each liquid may be considered to have a π–V curve of the general form shown for ether (curve 28). Arrows indicate the liquid molar volume at 20°C and atmospheric pressure; the position of the experimentally accessible part of the curve is determined by the properties of the particular liquid. For many non-polar organic liquids at room temperature the pressure range 1–2000 bar covers a broad maximum in the internal pressure–volume curve. On the other hand, in the higher density and incompressible liquids such as mercury (curve 1) and ionic liquids (molten salts) in curves 2–15 as a result of the strong cohesive coulombic forces, the $\pi-V$ curves are steeper and a smaller proportion of the typical complete curve is obtained for the same pressure range.

Table 2.1   Key to Fig. 2.5, internal pressure–volume data
(after Barton, 1971)

	Liquid	Temperature range (°C unless specified)	Pressure range (kbar unless specified)
1	Hg	20 – 50	0 –13
2	LiNO$_3$	m.p.– 430	0 –1.3
3	NaNO$_3$	m.p.– 460	0 –1.5
4	KNO$_3$	m.p.– 460	0 –1.5
5	RbNO$_3$	m.p.– 510	0 –1.2
6	CsNO$_3$	m.p.– 520	0 –1.2
7	LiCl	m.p.–1000	*
8	NaCl	m.p.–1000	*
9	NaBr	m.p.–1000	*
10	KCl	m.p.–1000	*
11	NaI	m.p.–1000	*
12	KBr	m.p.–1000	*
13	CsCl	m.p.–1000	*
14	CsBr	m.p.–1000	*
15	KI	m.p.– 800	*
16	A	84 – 140 K	–
		100 – 200 K	0.2–4
17	O$_2$	60 – 90 K	–
18	CH$_4$	90 – 130 K	–
19	N$_2$	63 – 120 K	–
20	H$_2$O (a) 25, (b) 45, (c) 85		0 –1
21	H$_2$O	200 – 850	1 –6
22	ethylene glycol (a) 25, (b) 105		0 –1
23	dimethylsulphoxide	15 – 37	0 –250 bar
24	benzene	(a) 125, (b) 65	0 –1
25	aniline	(a) 125, (b) 85	0 –1
26	nitrobenzene	(a) 25, (b) 85	0 –1
27	carbon tetrachloride	(a) 25, (b) 65	0 –1
28	ether	20 – 80	0 –12
		0 – 180	–
29	*n*-pentane	0 – 95	0 –10
30	*n*-hexane	10 – 50	1 –65 bar
		0 – 60	0 –5
31	*n*-heptane	20 – 36	1 –24 bar
		0 – 250	–
32	*n*-octane	0 – 60	0 –5
33	perfluoro-*n*-hexane	10 – 50	1 –65 bar
34	perfluoro-*n*-heptane	14 – 42	1 –60 bar
35	cyclohexane	25 – 95	1 –30 bar

18   * Calculated from atmospheric pressure ultrasonic measurements.

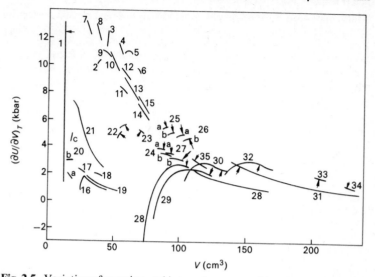

Fig. 2.5 Variation of experimental internal pressure with volume for a variety of liquids, including temperature dependence for those liquids in which this effect is important. The numbers refer to Table 2.1, and arrows indicate the liquid molar volume at 20°C and atmospheric pressure (after Barton, 1971).

It must be emphasized that for many liquids the internal pressure is dependent on temperature at constant volume; for example benzene (curve 24) and substituted benzenes (curves 25 and 26). Water at ambient conditions is shown in the curves labelled 20a, b, c for 25°, 45° and 85°C. This effect is attributed to the open, hydrogen bonded structure in water under ordinary conditions. A similar but less marked effect occurs in ethylene glycol (curve 22). However, for water at high pressures and high temperatures, the structure is broken down and the isochores approach linearity, i.e. the internal pressure becomes dependent on volume only (curve 21).

The effect of the internal pressure of a liquid on the spectra of weak chemical complexes has been discussed recently by Trotter (1966).

The tensile strength of liquids

An interesting demonstration of the strength of the cohesive forces in liquids is the existence of regions of tension ("negative pressure") under suitable conditions (Hamann, 1957; Hayward, 1964, 1970, 1971; Trevena, 1967; Apfel, 1972). The experimental limiting "negative pressures" of water (−277 bar), mercury (−425 bar) and organic liquids have been measured by a centrifugal method using filled capillaries open at both ends and rotating in a horizontal plane. Observations were made along the rotation axis to determine when the liquid column was broken by the force acting outwards in the liquid at each end of the capillary. Other techniques have involved the use of a pressure vessel. Recently measurements have been made on the $p–V–T$ properties of water under tension by the centrifugal method, and as might be expected, the constants in the empirical equation used to fit these data were very nearly those which described the positive pressure region (Winnick and Cho, 1971).

General references

⋆ BARKER, J. A. and HENDERSON, D. (1968), "The Equation of State of Simple Liquids", *J. Chem. Educ.*, 45, 2–6.

⋆ BARTON, A. F. M. (1971), "Internal Pressure: A Fundamental Liquid Property", *J. Chem. Educ.*, 48, 156–62.

⋆⋆ CAGLE, F. W. (1972), "A Classification of Equations of State", *J. Chem. Educ.*, 49, 345–7.

⋆⋆ DAYANTIS, J. (1972), "Considérations sur l'Équation de Tait and son Obtiention à partir de l'Équation de Van der Waals", *J. Chim. phys.*, 94–9.

⋆ FLOWERS, B. H. and MENDOZA, E. (1970), *Properties of Matter*, Wiley, London, particularly pp. 183–5.

⋆⋆ HAWARD, R. N. (1966), "Modified van der Waals Equation for Liquids", *Trans. Faraday Soc.*, 62, 828–37.

⋆⋆ HILDEBRAND, J. H. and SCOTT, R. L. (1950), *Solubility of Non-Electrolytes*, Reinhold, New York, 3rd edn.

⋆⋆ HILDEBRAND, J. H. and SCOTT, R. L. (1962), *Regular Solutions*, Prentice-Hall, Englewood Cliffs, New Jersey.

⋆⋆ HILDEBRAND, J. H., PRAUSNITZ, J. M. and SCOTT, R. L. (1970), *Regular and Related Solutions*, Van Nostrand Reinhold, New York.
These three books by Hildebrand and colleagues discuss the properties of non-electrolyte solutions, emphasizing the concept of solubility parameters and relating this to the liquid internal pressure.

⋆⋆ PRYDE, J. A. (1966), *The Liquid State*, Hutchinson, London, Chap. 2.

⋆⋆ RIGBY, M. (1970), "The van der Waals Fluid: A Renaissance", *Quart. Rev.*, 24, 416–32.

⋆⋆⋆ ROWLINSON, J. S. (1965), "The Equation of State of Dense Systems", *Rpts. Prog. Phys.*, 28, 169–99.

** ROWLINSON, J. S. (1969), *Liquids and Liquid Mixtures,* Butterworths, London, 2nd ed., Chap. 2.
An outline of the thermodynamic description of the liquid state.

* ROWLINSON, J. S. (1970), "Thermodynamics", *Chem. in Britain,* 6, 525–8.

** SCOTT, R. L. (1971), "Introduction" in *Physical Chemistry: An Advanced Treatise,* eds. Eyring, H., Henderson, D. and Jost, W. Vol. 8A, *Liquid State,* ed. Henderson, D. Academic Press, New York, Chap. 1, pp. 1–83.
A useful discussion at a reasonable level of many aspects of the description of liquid state properties.

** WINTERTON, R. H. S. (1970), "Van der Waals Forces", *Contemp. Phys.,* 11, 559–74.

Specific references

ALDER, B. J. (1972), "Correction to the van der Waals Model", *Physica,* 61, 152–6.

APFEL, R. E. (1972), "The Tensile Strength of Liquids", *Scientific American,* 227 (6), 58–71.

BARTON, A. F. M., HILLS, G. J., FRAY, D. J. and TOMLINSON, J. W. (1970), "High Pressure Densities of Molten Alkali Metal Nitrates: Compressibilities of Sodium Nitrate and Potassium Nitrate", *High Temperatures, High Pressures,* 2, 437–52.

HAMANN, S. D. (1957), *Physico-Chemical Effects of Pressure,* Butterworths, London, pp. 216–18.

HAYWARD, A. T. J. (1964), "Measuring the Extensibility of Liquids", *Nature,* 202, 481.

HAYWARD, A. T. J. (1970), "New Laws for Liquids: Don't Snap, Stretch!", *New Scientist,* 45, 196–9.

HAYWARD, A. T. J. (1971), "Negative Pressure in Liquids: Can it be Harnessed to Serve Man?", *Amer. Scientist,* 59, 434–43.

HILDEBRAND, J. H. (1929), "Intermolecular Forces in Liquids", *Phys. Rev.,* 34, 984–93.

MACDONALD, J. R. (1966), "Some Simple Isothermal Equations of State", *Rev. Mod. Phys.,* 38, 669–79.

MACDONALD, J. R. (1969), "Review of Some Experimental and Analytical Equations of State", *Rev. Mod. Phys.,* 41, 316–49.

MACDONALD, J. R. and POWELL, D. R. (1971), "Discrimination between Equations of State", *J. Res. Nat. Bur. Stand. A.,* 75, 441–53.

ROWLINSON, J. S. (1970), "Structure and Properties of Simple Liquids and Solutions: a Review", *Disc. Faraday Soc.,* 49, 30–42.

TREVENA, D. H. (1967), "The Behaviour of Liquids Under Tension", *Contemp. Phys.,* 8, 185–95.

TROTTER, P. J. (1966), "Spectra of Weak Chemical Complexes. Internal Compression Effect", *J. Amer. Chem. Soc.,* 88, 5721–6.

VAN DER WAALS, J. D. (1873), *On the Continuity of the Gas and Liquid Phases,* Dissertation, Leiden; English translation, Threlfall and Adair, *Physical Memoirs,* 1890, 1, 333.

WINNICK, J. and CHO, S. J. (1971), "*p–V–T* Behaviour of Water at Negative Pressures", *J. Chem. Phys.,* 55, 2092–7.

Chapter 3
Fluid mechanics and liquid transport coefficients

The characteristic property of liquids, distinguishing them from crystals, is fluidity. The fluidity (reciprocal of the viscosity) changes by many orders of magnitude on melting although the thermodynamic properties of liquids remain similar to those of the solid. Closely related to the fluidity are the other transport properties such as diffusion, electrical conductance and thermal conductance, and these are usually described in terms of fluid mechanics and macroscopic transport coefficients.

Viscosity

A liquid may be imagined as a set of xz planes stacked in the y direction and undergoing flow in the x direction such that a velocity

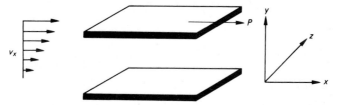

Fig. 3.1 Illustrating the definition of shear viscosity (equation (3.1)): laminar flow of a liquid between a fixed plate and a moving plate of unit area (adapted from Pryde, 1966).

gradient dv_x/dy results from an applied shear stress P_{xy} which is the force exerted on unit area of the xz plane in the x direction (Fig. 3.1). The shear viscosity η_s is defined by

$$P_{xy} = -\eta_s \frac{dv_x}{dy} \tag{3.1}$$

For a wide range of shear rates many liquids are Newtonian, i.e. η_s is independent of the rate of shear (and equal to the zero shear rate value), so η_s is a property which is characteristic of the liquid. This coefficient refers to fluid flows in which the fluid undergoes shear at constant volume. Another coefficient, called the second viscosity η', refers to the change of volume at zero shear. The SI unit for viscosity is the kg m^{-1} s^{-1} or pascal second, Pa s, and the now officially obsolete poise (P) is 0.1 Pa s.

Diffusion

The diffusion coefficient D_i of species i in a steady-state system at constant temperature and pressure is defined by Fick's first law, which is expressed

$$J_i = c_i v_i = -D_i(dc_i/dx) \tag{3.2}$$

for diffusion in the x direction. The flux J_i is the number of moles of i crossing unit area normal to x per second, c_i is the concentration in moles per unit volume and v_i is the average velocity of species i with respect to the system. The "driving force" for diffusion is the gradient of chemical potential, which for ideal liquid solutions is

$$(d\mu_i/dx) = -(RT/Lc_i)(dc_i/dx) \tag{3.3}$$

so the mobility u_i may be defined by

$$v_i = u_i(d\mu_i/dx) \tag{3.4}$$

or

$$D_i = u_i(RT/L) = u_i kT \tag{3.5}$$

In practice the diffusion coefficient usually measured is the tracer or self-diffusion coefficient where the "foreign" molecules are radioactive isotopes of the host molecules which are virtually indistinguishable from the latter as far as their dynamic behaviour is concerned.

Electrical conductance

The electrical conductivity κ is the ratio of the current I and the electric field strength E for unit volume of the sample contained between parallel electrodes separated by unit length. Ohm's law is followed by both electrolyte solutions and ionic liquids:

$$I = \kappa E$$
$$= (J_+ + J_-)F$$
$$= (c_+ v_+ - c_- v_-)F \tag{3.6}$$

Chapter 4
Liquid–gas continuity and corresponding states

For any substance the equation of state (Chapter 2) describes experimentally accessible information which can be depicted as a surface in p–V–T space. In some regions of this surface the substance can exist as a single phase, in other regions as two phases in equilibrium, and at one point (the triple point) as three phases in equilibrium. It is common practice to depict experimental p–V–T information in the form of perpendicular projections of this surface on to the p–T, T–V and p–V planes. Figures 4.1–4.3 show the form of such plots; these are diagrammatic and not to scale. The curves 1–4 in Fig. 4.3 and the lines 1–4 in Fig. 4.1 and 4.2 are isotherms, lines of constant temperature.

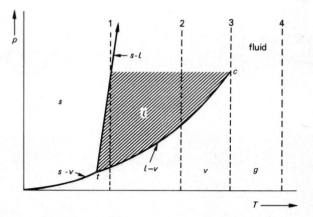

Fig. 4.1 p–T projection of a diagrammatic p–V–T three-dimensional phase diagram. The liquid phase regions is shaded (after Scott, 1971).

The liquid state (l) which is shaded, is defined by the vapour pressure curve (l–v in Fig. 4.1), the melting curve (s–l in Fig. 4.1), and the critical temperature (curve 3). The triple point (t) is the one p–V–T point where solid, liquid and gas are coexistent and here the density of the liquid is very close to that of the solid. At the critical point (c) the liquid and gaseous states are indistinguishable, so this

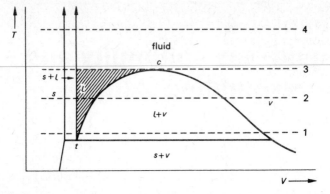

Fig. 4.2 *T–V* projection of a diagrammatic *p–V–T* three-dimensional phase diagram. The liquid phase region is shaded (after Scott, 1971).

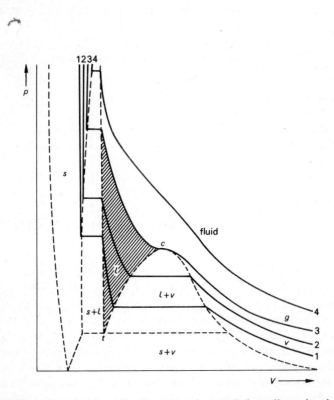

Fig. 4.3 *p–V* projection of a diagrammatic *p–V–T* three-dimensional phase diagram. The liquid phase region is shaded (adapted from Scott, 1971, and Chen, 1971).

30

(curve 3) marks the upper temperature limit at which the liquid state can exist. Here the density of the liquid and gaseous states are equal. It is noteworthy that the upper limit of the liquid state is set by the critical temperature, not the atmospheric pressure boiling point, and it is not the atmospheric pressure freezing point which is significant, but the triple point temperature. The freezing point and boiling point depend on the "accidental" factor of our atmospheric pressure, and unlike the critical and triple points have no universal significance. Even at atmospheric pressure, if a highly purified liquid is isolated from solid container surfaces it is possible to superheat and undercool it extensively. These limits for liquid water at atmospheric pressure are $-41°C$ to $279.5°C$, three times the familiar $0–100°C$ range (Apfel, 1972).

The vapour pressure curve terminates at the critical point, but no upper termination of the melting curve has been found experimentally – or is expected theoretically. Reference to Fig. 4.1 explains the term "liquid–gas continuity": it is possible to interconvert liquid and gaseous states without crossing a phase line by altering pressure and temperature in a suitable way. This might lead one to suppose that there is a closer similarity between liquids and dense gases than there is between liquids and solids. Although this may well be true, there are distinct differences, notably in the unique ability of liquids to form a free surface, which show that cohesion or attraction between molecules is very important in liquids.

The van der Waals equation of state

It is possible to test the van der Waals equation (2.6) by fixing the temperature and observing the isothermal variations of pressure with volume (Fig. 4.4). At small values of V, the isotherms approach the vertical line $V = Lb$, and at high temperatures (large p, large V) the isotherms approximate to hyperbolae corresponding to the ideal gas law (2.5). Equation (2.6) at lower temperatures yields sinuous curves which represent an unstable situation, although a metastable portion ed is experimentally accessible in a supersaturated vapour and ab corresponds to a superheated liquid state. At the temperature T_c the maximum and minimum meet at a point which can be identified with the critical point, and at low temperatures the isotherms intersect the volume axis, predicting the "negative pressure" region (Chapter 2). Thus the van der Waals equation predicts some of the main, general features of experimental fluid $p–V–T$ surfaces.

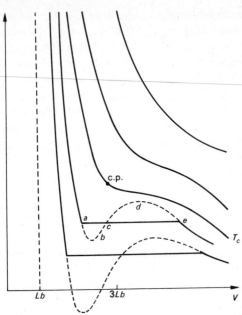

Fig. 4.4 Isotherms predicted by the van der Waals equation of state.

However, this simple equation cannot represent the discontinuities arising during the gas–liquid transition, nor can it predict the unusual dynamic properties (such as the critical opalescence mentioned at the end of Chapter 7) which occur in a fluid near its critical point. The behaviour of fluids near the liquid–gas critical point, particularly the transport or dynamic property behaviour is now receiving considerable attention (Stanley, 1971; Stanley *et al.*, 1971; Stephenson, 1971).

Reduced quantities and the law of corresponding states

In order to compare p–V–T surfaces for a variety of fluids, it is necessary to use p, V and T scales in which allowance is made for individual molecular properties. Because the critical quantities are considered to be more fundamentally important than other properties, it is customary to calculate reduced properties as the ratio of a given pressure, volume or temperature to the critical pressure, critical volume, or critical temperature:

$$p_r = p/p_c, \quad V_r = V/V_c, \quad T_r = T/T_c \tag{4.1}$$

32

It is then found for the simpler fluids that reduced isotherms nearly coincide, and this is particularly true where molecules have a similar chemical nature, e.g. the noble gases or a group of non-polar organic compounds. So if it is observed that a number of fluids have the same value for any two of p_r, V_r, T_r, then the third reduced quantity will also have very nearly the same value. This is the law of corresponding states.

A two-parameter equation of state such as the van der Waals equation can be written in the reduced form, that is in terms of p_r, V_r and T_r with no variable parameters. The reduced van der Waals equation may be obtained as follows. Writing equation (2.6) in order of descending power of V,

$$V^3 - V^2(Lb + RT/p) + aL^2V/p - abL^3/p = 0 \qquad (4.2)$$

and comparing with the general cubic equation at the critical point where all the roots are equal,

$$(V - V_c)^3 = V^3 - 3V^2V_c + 3VV_c^2 - V_c^3 = 0 \qquad (4.3)$$

it follows by comparing coefficients of like powers of V that

$$p_c = a/27b^2, \qquad V_c = 3bL, \qquad T_c = 8aL/27bR \qquad (4.4)$$

Substituting for a, b and R in equation (2.6) yields the reduced van der Waals equation,

$$(p_r + 3/V_r^2)(3V_r - 1) = 8T_r \qquad (4.5)$$

which contains no constant characteristic of a particular fluid.

Another simple way of testing the van der Waals equation is to evaluate RT_c/p_cV_c which from equation (4.4) should be $8/3 = 2.67$. Invariably in real fluids it is greater than this, often exceeding 4. For non-polar, approximately spherical molecule liquids it is ~ 3.5.

Of course, if an equation of state has more than three parameters, these cannot be all eliminated since there are only three experimental equations available (those relating p and p_r, V and V_r, T and T_r). In fact no three-parameter equation can represent precisely the p–V–T properties for all fluids, thus indicating the over simplified basis of this approach.

Reduced equations of state may be expressed alternatively in terms of the "force parameters", ε and σ, which describe molecular rather than bulk properties (Rowlinson, 1969; Scott, 1971). These molecular parameters are introduced in Chapter 11. The application of the corresponding states principle to the quantitative prediction of the physical properties of fluids has been reviewed recently by Leland and Chappelear (1968).

General references

** CRUICKSHANK, A. J. B. (1971), "Nonelectrolyte Liquids and Solutions", in *Problems in Thermodynamics and Statistical Physics,* ed. Landsberg, P. T., Pion, London, pp. 140–92.

** CHEN, S.-H. (1971), "Structure of Liquids", in *Physical Chemistry: An Advanced Treatise,* eds. Eyring, H., Henderson, D. and Jost, W., Vol. 8A, ed. Henderson, D. Academic Press, New York, Chap. 2 pp. 85–156, particularly pp. 85–7.

* PRYDE, J. A. (1966), *The Liquid State,* Hutchinson, London, Chaps. 1, 2.

*** ROWLINSON, J. S. (1969), *Liquids and Liquid Mixtures,* Butterworths, London, Chap. 8.

** SCOTT, R. L. (1971), "Introduction", in *Physical Chemistry: An Advanced Treatise,* eds. Eyring, H., Henderson, D. and Jost, W., Vol. 8A, ed. Henderson, D., Academic Press, New York, Chap. 1, pp. 1–83, particularly, pp. 1–8.

*** STANLEY, H. E. (1971), *Introduction to Phase Transitions and Critical Phenomena,* Clarendon Press, Oxford.
This monograph includes liquid–gas critical points together with ferromagnetic critical points in its treatment of the interdisciplinary field of phase transitions and critical phenomena. Chapter 4 deals with the van der Waals theory of liquid–gas phase transitions.

Specific references

APFEL, R. E. (1972), "Water Superheated to 279.5°C at Atmospheric Pressure", *Nature Phys. Sci.,* 238, 63–4.

LELAND, T. W. and CHAPPELEAR, P. S. (1968), "The Corresponding States Principle: A Review of Current Theory and Practice", *Ind. Eng. Chem.,* 60, No. 7, 15–43.

STANLEY, H. E., PAUL, G. and MILOŠEVIĆ, S. (1971). "Dynamic Critical Phenomena in Fluid Systems", in *Physical Chemistry: An*

Advanced Treatise, eds. Eyring, H., Henderson, D. and Jost, W., Academic Press, New York, Vol. 8B, ed. Henderson, D., Chap. 11, pp. 795–878.

STEPHENSON, J. (1971), "Critical Phenomena: Static Aspects", ibid., Chap. 10, pp. 717–93.

Chapter 5
Molecular distribution and correlation functions

Distribution functions in space

In order to discuss the nature of liquids in detail it is necessary to have a way of describing the environment of a representative molecule. A *molecular distribution function* provides a way of doing this in terms of a function averaged both over time and over all molecules, i.e. it provides an indication of the environment of a molecule smoothed in both space and time. (It does not, however, give a "snapshot" of a molecular distribution at a particular instant in time.)

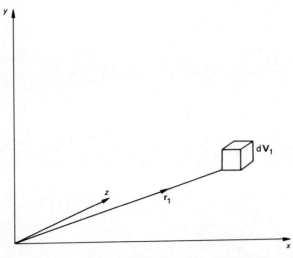

Fig. 5.1 Vector $\mathbf{r}_1$ defining a volume element dV_1.

There are three useful molecular distribution functions labelled $n^{(h)}$, where $h = 1, 2$ or 3. These will be introduced in terms of a vector diagram, that is a diagram in which both the direction and the magnitude (r) of the quantity $\mathbf{r}$ are important. Such a vector can describe the position of a small element of volume, which is then described as dV. The probability of finding one molecule in an element of volume dV_1 specified by vector $\mathbf{r}_1$ (Fig. 5.1) is $n^{(1)}(\mathbf{r}_1)\,dV_1$, where $n^{(1)}(\mathbf{r}_1)$ or simply $n^{(1)}$ is the *singlet distribution function*. Since probability is a pure number and dV_1 is a volume, $n^{(1)}$ has the dimensions of reciprocal volume or number density, and in an isotropic fluid at equilibrium it must equal the mean number density for the liquid, ρ:

$$n^{(1)}(\mathbf{r}_1) \text{ or } n^{(1)} = \rho = L/V \tag{5.1}$$

where L is the Avogadro constant and V is the molar volume. In other words, $n^{(1)}$ is independent of $\mathbf{r}_1$.

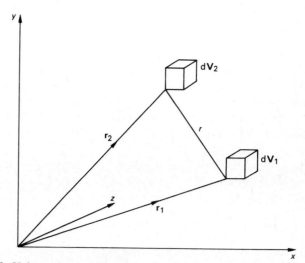

Fig. 5.2 Volume elements dV_1 and dV_2 defined by vectors $\mathbf{r}_1$ and $\mathbf{r}_2$ and with scalar separation r (after Rowlinson, 1970).

In discussing the *doublet distribution function* $n^{(2)}(\mathbf{r}_1, \mathbf{r}_2)$ or $n^{(2)}(r)$ or simply $n^{(2)}$ (Fig. 5.2), two elements of volume dV_1 and dV_2 with scalar separation $r = |\mathbf{r}_1 - \mathbf{r}_2|$ are considered. The probability of the simultaneous presence of molecules 1 and 2 in these volume elements is $n^{(2)}\,dV_1\,dV_2$, which is a function only of the scalar separation r and indepenent of the positions $\mathbf{r}_1$ and $\mathbf{r}_2$. It has proved useful to have a dimensionless doublet distribution function normalized to unity at large values of r, and the *pair distribution function* $g^{(2)}(r)$ is therefore defined by the equation

$$n^{(2)}(\mathbf{r}_1, \mathbf{r}_2) = n^{(1)}(\mathbf{r}_1)n^{(1)}(\mathbf{r}_2)g^{(2)}(r) \tag{5.2}$$

From equation (5.1),

$$n^{(2)}(\mathbf{r}_1, \mathbf{r}_2) = \rho^2 g^{(2)}(r),$$

and as $r \to \infty$, so

$$g^{(2)}(r) \to 1 \qquad (5.3)$$

because if r is large the probability of occupancy of volume element 2 is independent of whether or not volume element 1 is occupied, i.e. the probability of simultaneous occupancy is the product of independent occupancies and there is no long-range order or correlation between the two molecules.

The pair distribution function may also be defined

$$g^{(2)}(r) = \frac{1}{\rho^2}\langle \rho(0)\rho(\mathbf{r}) \rangle \qquad (5.4)$$

describing how the density with a value $\rho(0)$ at the origin changes with distance from the origin $\mathbf{r}$. The angular brackets denote an average over all initial positions $\mathbf{r} = 0$. Again it can be seen that as $|\mathbf{r}|$ or $r \to \infty$, $g^{(2)}(r) \to 1$ because at large distances the two local densities at 0 and $\mathbf{r}$ are statistically independent and the mean of their products is simply ρ^2.

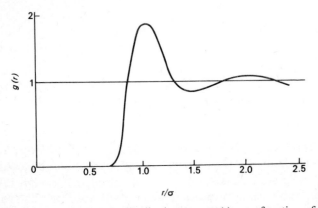

Fig. 5.3 Diagrammatic pair distribution curve $g(r)$ as a function of inter-molecular separation r. σ is an intermolecular separation parameter (Chapter 11).

It can be seen from a typical pair distribution curve (Fig. 5.3) that in a fluid $g^{(2)}(r)$ goes to zero as r goes to zero because two molecules cannot be closer than their diameters allow. After a series of maxima and minima, $g^{(2)}(r)$ approaches the value of unity, corresponding to complete disorder. The way in which pair distribution functions are experimentally determined from radiation scattering will be discussed in Chapters 6 and 7.

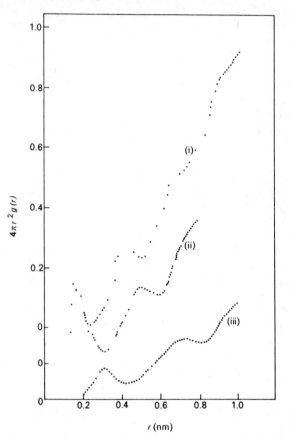

Fig. 5.4 Radial distribution function $4\pi r^2 g(r)$ for three liquid metals. (i) Liquid copper at 1313 K; (ii) liquid gold at 1373 K; (iii) liquid sodium at 373 K. Vertical scales are successively shifted by 10 units (after Bagchi, 1972).

The function $g^{(2)}(r)$ is sometimes called the radial distribution function, but the description "radial" should be strictly reserved for the quantity $4\pi r^2 g^{(2)}(r)$ (Fig. 5.4). A "running coordination" number, n, may be evaluated from $g^{(2)}(r)$ (Fig. 5.5). It is the average number of neighbours *within* a radial distance s of a reference molecule, in contrast to $g^{(2)}(r)$ which is the average number at a

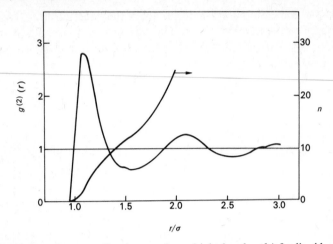

Fig. 5.5 Running co-ordination number n (right-hand scale) for liquid argon from a pair distribution function $g^{(2)}(r)$ (left-hand scale) computed by a molecular dynamics method (Chapter 12) (after Rahman and Stillinger, 1971).

particular r, normalized to 1 at large distances. The running co-ordination number is defined as $n = 4\pi\rho\int_{o'}^{s}r^2 g^{(2)}(r)\,dr$. The first coordination number (effectively the number of nearest neighbours) may be estimated by evaluating the area under the first peak of Fig. 5.5. For liquid argon this varies from ~ 10 at the triple point (compared with 12 in solid argon) to ~ 4 at the critical density.

Although there is usually a volume *increase* when a crystal melts, it is found from radiation scattering experiments that the molecules making up the liquid are on average the same distance apart as in the solid, or even closer. This apparent contradiction is explained if empty spaces occur in the liquid (see Fig. 10.1), and this explanation is supported by the much greater mobility of molecules in a liquid. The problem is to explain how this "empty space" is distributed.

It is possible to analyse in more detail liquid state pair distribution functions by comparing them with the crystalline state results, but it is important to remember that this is on the basis of purely spatial information which neglects the discussion of time. If this detailed analysis is done (Bagchi, 1970, 1972) the liquid is shown to consist of clusters of disordered microcrystallites with linear dimensions estimated at 5–15 nm (50–150Å). Thus the liquid state must be described by a statistical thermodynamic treatment of "small" systems, intermediate between molecules and bulk systems. It is suggested that crystals and gases have homogeneous distributions for probability of occurrence of neighbours, whereas liquids have inhomogeneous distributions. Thus the first-order phase transitions between crystals, liquids and gases correspond

molecule within a "cage" formed by its neighbours, resulting in frequent reversals of velocity.

The concept of correlation functions will be extended in the next chapter to introduce spectroscopic techniques for liquid structure investigation.

Distribution functions and correlation functions

Confusion sometimes arises in the terminology of distribution and correlation functions. In general *distribution functions* at large particle separations are normalized to unity (like $g^{(2)}(r)$) or tend to bulk values (like $G(\mathbf{r}, t)$ which tends to the bulk density). The term *correlation function* is used for functions which tend to zero when the volume elements under consideration are far apart, i.e. when there is no correlation. An example is the space–time correlation function $G'(\mathbf{r}, t)$.

The total correlation function and direct correlation function

It is frequently convenient to use a correlation function in place of the pair distribution function. The total correlation function is simply defined as

$$h(r) = g^{(2)}(r) - 1 \qquad (5.10)$$

and a typical curve is depicted in Fig. 5.8b, with the corresponding pair distribution function in Fig. 5.8a for comparison.

The correlation function $h(r)$ is a measure of the *total* influence of molecule 1 on molecule 2 at a distance $r = |\mathbf{r}_1 - \mathbf{r}_2|$, and (by definition) may be divided into two parts: a *direct correlation function*, $c(r)$, describing the direct effect of molecule 1 on molecule 2, and an indirect part which acts through the influence of other molecules. The function $c(r)$ is defined by the equation

$$h(\mathbf{r}_1, \mathbf{r}_2) = c(\mathbf{r}_1, \mathbf{r}_2) + \rho \int c(\mathbf{r}_1, \mathbf{r}_3) h(\mathbf{r}_2, \mathbf{r}_3)\, d\mathbf{r}_3 \qquad (5.11)$$

The full indirect effect is the sum of the separate effects involving the $(L\text{-}2)$ molecules other than 1 and 2, averaged over all the positions available in the volume V. The direct correlation function $c(r)$ acts over a much shorter range than $h(r)$ (see Fig. 5.8c) and therefore is more simply related to molecular properties such as molecular potential energy functions (Chapter 11). The reason for the form of

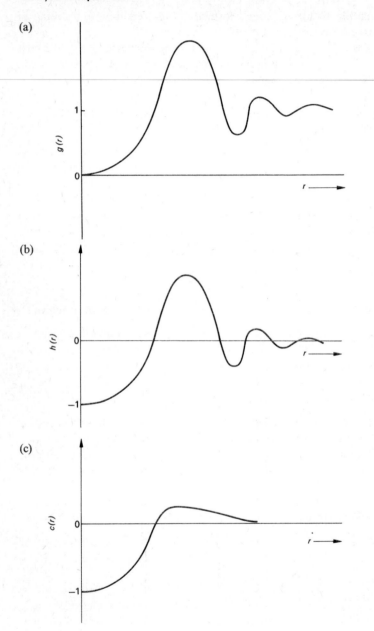

Fig. 5.8 Schematic diagram of (a) pair distribution function $g^{(2)}(r)$; (b) total correlation function $h(r)$; and (c) direct correlation function $c(r)$ to the same scale.

equation (5.11) may be seen more clearly by repeated expansions of $h(r)$ as $[c(r) +$ integral]:

$$h(\mathbf{r}_1, \mathbf{r}_2) = c(\mathbf{r}_1, \mathbf{r}_2) + \rho \int c(\mathbf{r}_1, \mathbf{r}_3) c(\mathbf{r}_3, \mathbf{r}_2)\, d\mathbf{r}_3 +$$
$$\rho^2 \iint c(\mathbf{r}_1, \mathbf{r}_3) c(\mathbf{r}_3, \mathbf{r}_4) c(\mathbf{r}_4, \mathbf{r}_2)\, d\mathbf{r}_3\, d\mathbf{r}_4 + \cdots \tag{5.12}$$

In this way $h(\mathbf{r}_1, \mathbf{r}_2)$ is split up into the direct interaction between molecules 1 and 2, and the indirect effect through all possible chains of direct correlations.

General references

★ BARTON, A. F. M. (1972), "The Description of Dynamic Liquid Structure", *J. Chem. Educ.*, 50, 119–22.

★★ BAXTER, R. J. (1971), "Distribution Functions", in *Physical Chemistry: An Advanced Treatise,* eds. Eyring, H., Henderson, D. and Jost, W., Vol. 8A, *Liquid State,* ed. Henderson, D., Academic Press, New York, Chap. 4, pp. 267–334.

★★ EGELSTAFF, P. A. (1967), *An Introduction to the Liquid State,* Academic Press, London, Chaps. 2, 8.

★ ROWLINSON, J. S. (1970), "The Structure of Liquids", in *Essays in Chemistry,* eds. Bradley, J. N., Gillard, R. D. and Hudson, R. F., Academic Press, London, Vol. 1, pp. 1–24.

★★ SCHOFIELD, P. (1968), "Experimental Knowledge of Correlation Functions in Simple Liquids", in *Physics of Simple Liquids,* eds. Temperley, H. N. V., Rowlinson, J. S. and Rushbrooke, G. S., North-Holland, Amsterdam, Chap. 13, pp. 563–609.

Specific references

BAGCHI, S. N. (1970), "Kinematic Theory of Diffraction by Matter of any Kind and the Theory of Liquids", *Adv. Phys.,* 19, 119–73.

BAGCHI, S. N. (1972), "The Structure of Liquids as Revealed by the Analysis of their Radial Distribution Functions", *Acta Cryst.,* A28, 560–71.

EGELSTAFF, P. A., PAGE, D. I. and HEARD, C. R. T. (1971), "Experimental Study of the Triplet Correlation Function for Simple Liquids", *J. Phys. (C),* 4, 1453–65.

RAHMAN, A. (1964), "Correlations in the Motions of Atoms in Liquid Argon", *Phys. Rev.,* 136, A405–11.

RAHMAN, A. and STILLINGER, F. H. (1971), "Molecular Dynamics Study of Liquid Water", *J. Chem. Phys.,* 55, 3336–59.

VAN HOVE, L. (1954), "Correlations in Space and Time and Born Approximation Scattering in Systems of Interacting Particles", *Phys. Rev.,* 95, 249–62.

Chapter 6
Spectroscopic techniques for liquid dynamic structure studies

Many spectroscopic techniques are available for the study of inter-molecular motions in liquids. The shapes of spectral peaks or bands in techniques such as infra-red absorption and Raman and inelastic neutron scattering (which are widely used to study *intra*molecular properties for analysis or molecular structure determination) contain information on *inter*molecular properties. Other methods such as dielectric, acoustic and magnetic relaxation also contribute to our knowledge of liquid dynamics.

All these techniques use as a "probe" an external field which is *weakly* coupled to the system, and study the response of the liquid system to the probe. The weak nature of the coupling ensures that the probe does not influence or obscure the dynamical behaviour of the liquid. In these circumstances the response may be described in terms of the time correlation functions of the dynamic properties, introduced in the previous chapter.

Time correlation functions

In recent years, time correlation functions have been applied to all types of spectroscopic techniques. Particles such as neutrons or electromagnetic waves such as visible light interact with molecules because of different molecular properties. Thus neutrons are scattered by nuclei, and infra-red radiation is transferred when the dipole moment changes during vibration. As the positions of the nuclei and the orientation of the dipoles change with time, the scattered neutron beam or absorbed or scattered radiation responds to these changes. For example, in infra-red absorption spectroscopy the time correlation function for dipole moment, $\langle \mathbf{p}(0) \cdot \mathbf{p}(t) \rangle$ describes how a vibrating dipole orientated in a certain direction in the liquid at time $t = 0$ reorientates until as $t \rightarrow \infty$, $\langle \mathbf{p}(0) \cdot \mathbf{p}(t) \rangle \rightarrow 0$, i.e. until there is no correlation between its present and original positions. In other words, the correlation function describes how the dipole "forgets" its original orientation. The results of several different types of experiment, when combined, clarify the underlying dynamic processes in the liquid. Thus while infra-red absorption provides information about the dipole correlation function, Raman scattering is determined by the correlation function $\langle \alpha(0) \cdot \alpha(t) \rangle$ of the polarizability α, and neutron diffraction from molecular liquids reflects a complex combination of molecular properties which is frequently difficult to disentangle.

The experiments which detect molecular motions may be made directly in the "time domain"; that is, a property may be studied as a function of time, for example by applying a sharp electric field pulse and recording its decay on a fast response oscilloscope. More commonly, however, experiments are made in the "frequency domain", where the spectrum of frequencies ("line shape") is studied. In this case, if the frequency-dependent property (e.g. infra-red absorption) is measured over a wide *frequency* range, then in principle it can be transformed into the appropriate *time* correlation function. The way in which this transformation is done will be discussed later, but it is sufficient at present to appreciate that time-dependent properties and frequency-dependent properties are related to each other by a mathematical transformation process.

It will be shown below that a similar transformation property exists between the *momentum* transfer (i.e. the angle of scatter) when radiation interacts with matter, and interparticle *distances*. This idea is probably already familiar from the crystalline state X-ray diffraction pattern.

"Mechanical" or "viscoelastic" methods using acoustic (sound) energy, both shear and longitudinal techniques, differ from all other methods which involve energy in the form of radiation (either electromagnetic wave or neutron). This account is concerned mainly with radiation spectroscopy.

Absorption and scattering spectroscopy

Radiation spectroscopy techniques can be classified as "absorption spectroscopy" when quanta of radiation energy are completely absorbed, and "scattering spectroscopy" when they are not completely absorbed. When energy is imparted to or received from molecules by radiation, the scattering and absorption processes can occur either with the quantized motions of the nuclei (molecular vibrations and rotations, with discrete energy steps) and electronic transitions, or in the liquid state with non-quantized translational and rotational diffusion motion which provides information on the distribution of particles as a function of time and distance.

The optical absorption spectroscopy techniques are well known: absorption of visible and ultra-violet light is caused by electronic transitions; infra-red absorption corresponds to vibrational processes, and far infra-red and microwave spectroscopy provides information on particle rotation. In these absorption techniques, light quanta or photons of the appropriate energy (frequency) are completely absorbed.

However, a process may take only a portion of the energy from the incident radiation during a scattering process. For example, rotational "fine structure" or broadening is observed in vibrational spectra, corresponding to small shifts in the higher energy (frequency) radiation caused by the lower energy process. Therefore in general it is possible to observe a process with a particular energy or frequency either by studying absorption of energy of that frequency (absorption spectroscopy) or by studying the shift in energy of radiation of a different energy or frequency (scattering spectroscopy).

All radiation scattering spectroscopy techniques (e.g. neutrons, X-rays, light) have much in common: essentially monoenergetic radiation interacts with the molecules, and during the scattering process energy is exchanged so that a spectrum of energies is produced. A familiar example is vibrational Raman spectroscopy, which provides information on the infra-red and microwave frequency scale from shifts in light in the visible frequency region. In the same way, peaks in "inelastic" (i.e. accompanied by energy transfer) neutron scattering spectra can be identified with particular vibrational modes. *Quantized* motion yields *inelastic* neutron scattering or *Raman* light scattering, while non-quantized, diffusional motion gives rise to *quasi-elastic* neutron scattering in liquids (elastic scattering would take place for a perfectly rigid scattering centre in a solid) or *Brillouin* light scattering.

Quantized processes are associated with *intra*molecular properties (those within the molecule) and are therefore not of major interest in the study of the liquid state which is controlled by *inter*molecular processes (those between molecules). However, fine structure on the frequency unshifted (Rayleigh) and shifted (vibrational and rotational Raman) quantized motion spectral peaks arises from *inter*molecular diffusional scattering processes and is therefore of great importance.

Table 6.1 Intermolecular and intramolecular spectroscopy

	Absorption technique	Process	Scattering technique
Intermolecular, translational and rotational processes		non-propagating density fluctuations	Rayleigh
	[ultrasonic hypersonic]	propagating density fluctuations (sound waves)	Brillouin (Doppler effect)
	a.c. methods, audio, radiofrequency and microwave (Debye) dielectric relaxation	uncorrelated, random orientation — dipole moment orientation | polarizability change	depolarized Rayleigh (rotational Raman)
	far infra-red (Poley) absorption	correlated orientation (libration) rotation of dipole (permanent or collisionally induced dipole moment) — polarizability change	depolarized Rayleigh "collisional wings"
Intramolecular, quantized processes	infra-red	vibration — dipole moment change | polarizability change	vibrational Raman
	visible and ultra-violet	electronic transitions	
	X-ray	electronic transitions, (X-ray energy $\gg$ bonding energy)	
	γ-ray	Mössbauer	

Table 6.1 is included so that it may be referred to when a particular technique is mentioned; it indicates only the broad pattern of behaviour for liquids.

Electromagnetic radiation and neutrons

With electromagnetic radiation the excitation is *indirect*, via the electrons (and therefore in the case of quantized processes it is subject to selection rules imposed by dipole and induced dipole electronic properties) whereas neutron excitation occurs by direct impact of the neutron and the scattering nucleus. Another major difference between photon and neutron spectroscopy occurs in the widely different time scales exhibited. For a photon with its high velocity the characteristic time of interaction with a particle is $\sim 10^{-18}$ s. In condensed materials the shortest times of interest in connection with intermolecular behaviour are $\sim 10^{-14}$ s, so the particles as seen by each photon are effectively frozen. In the case of X-rays, energy changes in the photon due to changes in particle location during scattering are too small to be measured by present-day techniques. Different photons record different instantaneous configurations, and the pair distribution observed is "smeared" or averaged for the structure and does not provide time-dependent information (Kruh, 1962). Neutron scattering on the other hand can provide detail about the time-dependent liquid structure. Because the neutron velocity used is much lower than that of light, its interaction time with the particle during scattering is $\sim 10^{-11}$ s, and it will be shown how the Doppler broadening of the scattered neutrons provides information on the microdiffusion process.

In general plots of intensity versus radiation energy are known as absorption or scattering spectra, and plots of intensity versus angle are known as diffraction patterns. Scattering is a molecular or atomic process and the macroscopic effect of scattering interference is called diffraction. The scattering of neutrons gives rise to both scattering spectra (time or energy dependence information) and diffraction patterns (distance or momentum information).

The mathematical description of radiation scattering

The concept of X-ray diffraction in a crystal is well known: an array of atoms provides a periodic potential which scatters the incident wave and generates a diffraction pattern. The position is similar in neutron diffraction, except that in this case the scattering is done by nuclei rather than electrons, so the same general mathematical description can be applied to all forms of radiation. For atoms involved in vibrations or translations, as in a liquid, the scattering potential becomes time-dependent, and the scattered wave suffers angular frequency shifts – the "Doppler effect". The momentum transfer associated with this energy change is added to that from the solid-like diffraction process.

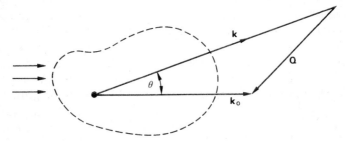

Fig. 6.1 Vector diagram of equation (6.1), $\mathbf{Q} = \mathbf{k}_0 - \mathbf{k}$ (after Chen, 1971).

Consider a scattering process in which the momentum transferred is $\hbar\mathbf{Q}$ (i.e. $h\mathbf{Q}/2\pi$), where $\mathbf{Q}$ is the difference in the wave vector between incident and scattered radiation (Fig. 6.1)

$$\mathbf{Q} = \mathbf{k}_0 - \mathbf{k} \tag{6.1}$$

In this equation, $\mathbf{k}_0$ and $\mathbf{k}$ are the incident and scattered wave vectors, k_0 and k being the corresponding scalar wavenumbers ($2\pi/\lambda$, where λ is the wavelength). This momentum transfer is associated with scattering through an angle θ, so

$$Q^2 = k^2 + k_0^2 - 2k\,k_0 \cos\theta \tag{6.2}$$

During the scattering process the energy transferred is given by

$$\delta E = E_0 - E = \hbar\omega \tag{6.3}$$

It can be seen that Q has dimensions of reciprocal distance (proportional to λ^{-1}) and ω has the dimensions of reciprocal time: in other words $1/Q$ is a distance and $1/\omega$ is a time. As indicated above, diffraction data on the extent of scattering as a function of momentum transfer $\hbar Q$ can be transformed into information on intermolecular distances, and data on the extent of scattering as a function of energy transfer $\hbar\omega$ can be transformed into information on the dependence of intermolecular distances on time. In order to obtain the most detailed and precise information on the time correlation functions in a liquid, it is necessary for the "interaction time" of the energy with the liquid, $\hbar/\delta E$, to be comparable to the time scale of molecular processes, and for the radiation wavelength λ ($=2\pi/k_0$) to be of the order of magnitude of intermolecular distances.

The neutron scattering technique best satisfies these requirements, $Q^{-1} \sim 10^{-10}$ m (1Å), $\omega^{-1} \sim 10^{-13}$ s, and it is therefore especially valuable and discussed in detail in the next chapter. Although in principle neutron scattering provides sufficient information on the structure and dynamics of liquids to describe all equilibrium and transport properties, practical difficulties prevent complete realization of its potential. Information from all techniques must therefore be combined in an attempt to understand the detailed behaviour of liquids. Figure 6.2 indicates typical radiation wavelengths and

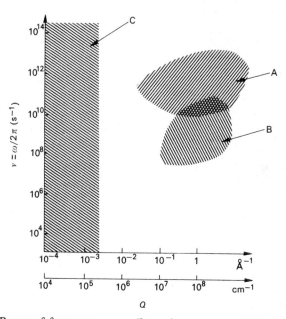

Fig. 6.2 Range of frequency $\nu = \omega/2\pi$ and wave vector Q accessible to scattering experiments. A, conventional neutron spectrometers; B, high resolution neutron spectrometer; C, optical spectroscopy (after Springer, 1972).

interaction times for several scattering techniques. Conventional neutron spectrometers have a resolution of $\delta E = 10^{-5}$–10^{-4} eV $(10^{-24}$–10^{-23} J) which corresponds to a time scale of 10^{-10}–10^{-11} s. A newly developed spectrometer has a resolution two orders of magnitude greater, corresponding to times as long as 10^{-8} s (Springer, 1972). Optical spectroscopy (Raman, infra-red, Brillouin, Rayleigh) covers a very wide frequency range, but is limited to low Q values. As Q^{-1} is several hundred times interatomic distances for visible light, these techniques provide information on long-range fluctuations. For X-rays Q^{-1} is of the order of interatomic spacing, but the energy change δE cannot be measured with current instruments. Only the Mössbauer technique, using gamma radiation, is sufficiently sensitive, but this has a rather limited application. The resonance absorption of gamma rays is determined by the self-correlation function $G_s(\mathbf{r}, t)$ describing the individual motion of the absorbing nucleus, and for an atom with some mobility the Mössbauer absorption peak is broadened by the diffusive motion if the natural nuclear life-time is comparable with the characteristic times of the motion. This corresponds to extremely slow diffusive motions, and so applies only to very viscous media (see, for example, Abras and Mullen, 1972). The relaxation methods of nuclear magnetic resonance, dielectric and ultrasonic measurements apply to a frequency range below 10^{10} Hz.

General references

★★ CHEN, S. H. (1971), "Structure of Liquids", in *Physical Chemistry: An Advanced Treatise,* eds. Eyring, H., Henderson, D. and Jost, W., Vol. 8A, *Liquid State,* ed. Henderson, D., Academic Press, New York, Chap. 2, pp. 85–156.
A general review of liquid structural properties with a considerable proportion devoted to radiation scattering. The mathematical description is more detailed than in this book, but a number of graphs of experimental results will be useful.

★★★ CHU, B. (1970), "Laser Light Scattering", *Ann. Rev. Phys. Chem.,* 21, 145–74.

★★★ EGELSTAFF, P. A. (1967a), *An Introduction to the Liquid State,* Academic Press, London, Chaps. 8, 9.

★★ EGELSTAFF, P. A. (1967b), "Radiation Scattering Studies of the Structure and Transport Properties of Liquids", *Disc. Faraday Soc.,* 43, 149–59.

★ EGELSTAFF, P. A. and SCHOFIELD, P. (1965), "The Structure and Thermal Motion of Simple Liquids, Parts I and II", *Contemp. Phys.,* 6, 274–84, 453–64.

★★ GORDON, R. G. (1968), "Correlation Functions for Molecular Motion", in *Advances in Magnetic Resonance,* ed. Waugh, J. S., Academic Press, New York, pp. 1–42.
A very valuable review of this subject at a moderately advanced level.

★★ PETICOLAS, W. L. (1972), "Inelastic Light Scattering and the Raman Effect", *Ann. Rev. Phys. Chem.,* 23, 93–116.

★★ WHITE, J. W. (1971), "Neutron Scattering Spectroscopy in Relation to Electromagnetic Methods", *Molecular Spectroscopy,* Institute of Petroleum, London, pp. 199–222.

Specific references

ABRAS, A. and MULLEN, J. G. (1972), "Mössbauer Study of Diffusion in Liquids: Dispersed Fe^{2+} in Glycerol and Aqueous–Glycerol Solutions", *Phys. Rev. (A),* 6, 2343–53.

KRUH, R. F. (1962), "Diffraction Studies of the Structure of Liquids", *Chem. Rev.,* 62, 319–46.

SPRINGER, T. (1972), "Quasielastic Neutron Scattering for the Investigation of Diffusive Motions in Solids and Liquids", *Springer Tracts in Modern Physics,* ed. Höhler, G., Vol. 64, Berlin–Heidelberg–New York: Springer.

Chapter 7
Neutron and X-ray scattering spectroscopy

The scattering of neutrons by diamagnetic materials arises from interactions with the nuclei; so neutron scattering yields information about the translational motion of nuclei, including the rotation of molecules. (In paramagnetic material, with unpaired electron spins, the scattering is mainly by magnetic interactions.) Because there is a weak coupling between neutrons and nuclei, the scattering can be described by a correlation function for translational motion, i.e. by a van Hove correlation function as described in Chapter 5.

In an atomic liquid the translational motion is due purely to the centre of mass motions, but in a molecular liquid part of the translational motion of the nuclei arises from rotation of the molecules. In general the rotational and translational motion are coupled together, so neutrons probe a complicated mixture of the two. The mathematical treatment in this chapter will for simplicity be restricted to *atomic* liquids. There are two main kinds of measurement which can be made. The first is analogous to X-ray diffraction from a crystal, which has a basically time-independent structure.

Correlation in position: the static structure factor

Experiments may be carried out in which no energy dependence is studied, but the diffracted radiation (the total intensity of radiation, integrated by the detector over all possible energy transfers δE) is recorded as a function of momentum transfer $\hbar Q$ (i.e. as a function of the angle of scatter) and the results (intensity versus Q) are described as the *static structure factor*, $S(Q)$. (In liquids $S(Q)$ is a function of the magnitude only of $\mathbf{Q}$, and is therefore written without vector notation for Q.) The static structure factor is plotted for liquid argon in Fig. 7.1 and for liquid lead in Fig. 7.2.

The static structure factor provides information on the time-independent, position-averaged local structure of the liquid, and is closely related to the pair distribution function $g^{(2)}(r)$. The main peak in $S(Q)$ reflects the existence of short-range order, also responsible for the oscillations in $g^{(2)}(r)$ (Fig. 5.3). The peak in $S(Q)$ for liquids

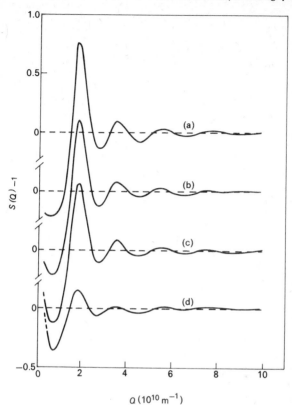

Fig. 7.1 $(S(Q)-1)$ for argon from X-ray scattering at $-125°C$. The densities are (a), 0.982 g cm^{-3}; (b), 0.910 g cm^{-3}; (c), 0.780 g cm^{-3} and (d), 0.280 g cm^{-3} (after Mikolaj and Pings, 1967).

corresponds to a diffraction "spot" in the X-ray diffraction pattern for a crystal, where information on the angle of scattered radiation (i.e. momentum transfer) can be transformed into information on atomic distances. The *position* of the $S(Q)$ peak (at Q_0) is determined by the *period* $(2\pi/Q_0)$ of the $g^{(2)}(r)$ oscillations. At small values of Q, which are strongly influenced by the liquid properties at large values of r, the $S(Q)$ value is determined by the isothermal compressibility κ_T, a macroscopic quantity:

$$\kappa_T = \frac{1}{\rho}\left(\frac{\partial \rho}{\partial p}\right)_T = \frac{1}{\rho k T}\lim_{Q\to 0} S(Q) \qquad (7.1)$$

The oscillatory behaviour of $S(Q)$ with successive peaks at multiple

59

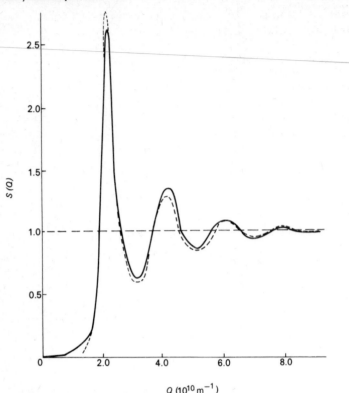

Fig. 7.2 $S(Q)$ for liquid lead by neutron (——) and X-ray (– – –) scattering at 340°C and 329°C respectively (after Enderby, 1968).

values of Q_0 corresponds to the sharp peak in $g^{(2)}(r)$ at $r = 2\pi/Q_0$. At large values of Q, $S(Q)$ is strongly damped, and it falls away asymptotically to unity. In fact, $S(Q)$ and $g^{(2)}(r)$ are related by a Fourier transformation (equation (A.8)). This mathematical technique is discussed in the Appendix although complete familiarity with this procedure is not essential because the effect can be seen by comparing the $S(Q)$ and $g^{(2)}(r)$ plots. To summarize:

  (i) the angular distribution of diffracted radiation intensity is the space Fourier transform of the pair distribution function;

 (ii) the strong peaks in $S(Q)$ correspond to oscillations in $g^{(2)}(r)$ and vice versa;

(iii) the low Q part of $S(Q)$ is influenced most strongly by the large r part of $g^{(2)}(r)$.

Comparison of neutron and X-ray diffraction for pair distribution functions

Pair distribution functions can be determined by either low energy neutron diffraction or X-ray diffraction. Similar mathematical descriptions are valid, but the terminology of neutron diffraction is used here. Neutron diffraction is preferable in fluids containing atoms with high atomic numbers which exhibit high X-ray absorption, but X-ray diffraction is usually considered preferable to neutron diffraction for the determination of $g^{(2)}(r)$ in other fluids because X-ray diffraction apparatus can be constructed to give higher angular resolution.

The dynamic structure factor

The second type of radiation scattering experiment is that in which the cross-section (the intensity of scattered radiation) is measured as a function both of the momentum transferred, $\hbar Q$ (the angular dependence), and the energy transferred, $\hbar \omega$. This more detailed information is described as the *dynamic structure factor*, $S(Q, \omega)$.

It is instructive to consider the relationship between $S(Q)$ and $S(Q, \omega)$ for the case of *X-ray* radiation. In this case,

$$k \approx k_0,$$

so substituting $k = k_0 = 2\pi/\lambda$ and the cosine formula into equation (6.2), it follows that

$$Q = \frac{4\pi \sin (\theta/2)}{\lambda} \tag{7.2}$$

This indicates that the momentum transfer $\hbar Q$ depends only on the incident wavenumber k_0, and is independent of the energy $\hbar \omega$ and wavenumber k of the scattered radiation. In this simple case, $S(Q, \omega)$ may be integrated over ω:

$$S(Q) = \int_{-\infty}^{+\infty} S(Q, \omega) \, d\omega \tag{7.3}$$

In *neutron* scattering the approximation $k \approx k_0$ cannot be made and the structure factor represents a more complex form of integration over the energy transfer. Nevertheless the same general principle holds: $S(Q)$ may be obtained from $S(Q, \omega)$ with a consequent loss of information, but $S(Q, \omega)$, of course, cannot be evaluated from $S(Q)$ data.

Experimental neutron scattering spectroscopy*

In practice, a beam of thermal neutrons, made almost mono-energetic by velocity selection, falls on the sample and the neutrons are scattered through various angles in collisions with the (moving) atomic nuclei. Thermal neutrons from the moderator of a reactor are passed through additional moderators and filters (for example, liquid hydrogen to reduce their energy to a Maxwell–Boltzmann distribution at ~ 30 K), then the beam passes on to a rotating chopper which pulses the beam and at the same time transmits only neutrons with the small range of velocities defined by the curvature of the slots in the rotor and by its spinning speed. The time of flight of the scattered neutrons from the sample to the detector provides a measure of their energy. Typically (Aldred *et al.*, 1967) the sample is inclined at $45°$ to the neutron beam and the boron trifluoride counters are placed at distances of ~ 1 m from the sample and at angles between 5 and $90°$ of inclination from the incident beam (Fig. 7.3).

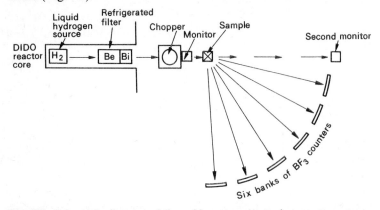

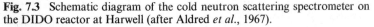

Fig. 7.3 Schematic diagram of the cold neutron scattering spectrometer on the DIDO reactor at Harwell (after Aldred *et al.*, 1967).

Since the angle of scatter cannot exceed $180°$, the maximum value of Q is $\sim 1.2 \times 10^9$ m^{-1} (0.12 Å^{-1}) if the wavelength is ~ 0.1 nm (1 Å). (For light scattering, it can be seen from equation (7.2) on inserting the minimum convenient wavelength, $\lambda = 200$ nm (2000 Å) that the maximum value of Q attainable is $\sim 6 \times 10^7$ m^{-1} (6×10^{-3} Å^{-1}).) If neutrons thermalized at liquid hydrogen temperature are used with room temperature scattering samples, the dominant inelastic scattering process is one in which the neutrons *gain* energy

* Aldred *et al.*, 1967; Egelstaff, 1965.

in collisions, i.e. the anti-Stokes scattering is more probable than the Stokes scattering.

The self-correlation function $G_s(\mathbf{r}, t)$ (equation (5.7)) describing the average motion of a single particle can be determined if the amplitude and phase of the scattered radiation varies from particle to particle, i.e. if the scattering is *incoherent*. The properties of incoherence or coherence in neutron scattering depend on details of the interaction in the scattering process between the neutron and the nucleus, and will not be discussed here. Most nuclei scatter both coherently and incoherently, but hydrogen is almost completely incoherent and also a very strongly scattering nucleus, permitting the determination of the self-correlation function in organic liquids. The ability of a nucleus to scatter neutrons is measured by the "cross-section", σ, the larger the cross-section the stronger being the scattering. Table 7.1 lists typical values for coherent and incoherent

Table 7.1 Typical neutron scattering cross-sections in barns (1 barn = 10^{-28} m^2) (after Springer, 1972)

Element	$\sigma_{coherent}$	$\sigma_{incoherent}$
H	1.8	80
D	5.4	2.2
C	5.5	negligible
O	4.2	negligible
Na	1.6	1.9
Ar	0.5	0.4

cross-sections. Experiments on incoherent scattering are the most useful, because although coherent scatterers give the most complete information, the interpretation of their spectra is very difficult. For molecules with chemically non-equivalent, incoherently scattering atoms the contribution from each type of atom can be treated separately and additively.

It is customary to plot spectra of intensity against energy (e.g. neutron velocity or time of flight) for a variety of momentum values (i.e. at different scattering angles), and so determine a surface in three-dimensional space with axes intensity, energy or ω, and momentum or Q (Fig. 7.4). At any Q value, the intensity–energy graph falls into two regions. In the *inelastic* region of the energy range (energy transfer associated with quantized vibrational modes in a molecular liquid and absent in Fig. 7.4 for the monatomic liquid argon) neutron scattering information is complementary to optical spectroscopic data. It is usually more difficult to assign

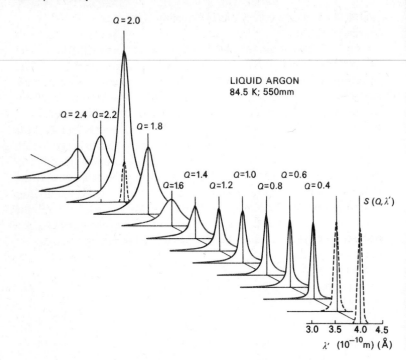

Fig. 7.4 The scattering surface for liquid argon: $S(Q, \omega)$ plotted as a function of wave vector transfer Q and the wavelength λ' of the outgoing neutrons (after Dasannacharya and Rao, 1965).

bands in the neutron spectrum to particular vibrations, but by making selective substitution of deuterium and fluorine for hydrogen atoms, for example, it is possible to reduce the neutron scattering cross-section of a particular group by at least an order of magnitude and so identify the contribution of that group to the neutron spectrum (Aldred *et al.*, 1972). This information concerns *intra*molecular properties, and precise details of rotational broadening are not yet available from this technique.

The other energy region is the *quasi-elastic* peak, the term "quasi-elastic" being used to describe the region of small energy transfers which arise when the neutron is scattered from the random molecular motions. Formally, the dynamic structure factor $S(Q, \omega)$ may be related to the van Hove correlation function $G'(\mathbf{r}, t)$ (equation (5.6)) in the same way that the pair distribution function $g^{(2)}(r)$ is related to the static structure factor $S(Q)$, but by a double Fourier transformation involving both t and ω, and $\mathbf{r}$ and Q (equation (A.9)). The self-correlation function $G_s(\mathbf{r}, t)$ (equation (5.7)) describes the motion that would be observed on average if the centre of a certain particle was followed as a function of time, and this is related by a double Fourier transform (equation (A.10)) to the self-dynamic structure factor $S_s(Q, \omega)$ obtained from incoherent neutron scattering. In general such complete transformations are not yet possible because of experimental limits on Q and ω, and the experimental data are treated in a variety of alternative ways to describe various aspects of liquid structure. The shape of the incoherent component in the quasi-elastic scattering may be discussed in terms of various models designed to combine the effects of centre of mass (translational) and rotational jump diffusion. This has been done, for example, by Winfield and Ross (1972) for C_6H_6 and C_6D_6. At the present state of development the (molecular) diffusion coefficient is the most useful piece of information obtained and this can be compared with the (bulk) tracer diffusion coefficient also available experimentally.

Comparison of neutron and optical spectroscopy

The self-dynamic scattering function $S_s(Q, \omega)$ obtained from incoherent neutron scattering experiments is a three-dimensional "line shape" of scattering as a function of momentum and energy. The analogous quantity for *infra-red absorption* and *Raman scattering* is a two-dimensional line shape which is a function of energy alone because the photon momentum transfer is negligible. (The photon momentum transfer although small is important in Brillouin scattering: see below.) Thus the neutron and optical spectra may be compared at the low Q limit of radiation scattering, although the optical spectra are Fourier transforms of the time correlation functions of dipole moment and polarizability respectively (see Chapter 9), rather than of the velocity correlation function which is appropriate for neutron scattering. The qualitative behaviour of $S(Q, \omega)$ is indicated in Fig. 7.5. As already shown, neutron scattering differs fundamentally from electromagnetic radiation spectroscopy in that it allows one to study high Q values, where $S(Q, \omega)$ becomes independent of Q and dependent only on ω. This is the "ideal gas" or "single particle" region (d). The other extreme is the low Q or cooperative region (a), and the $Q \to 0$ limit reveals macroscopic properties.

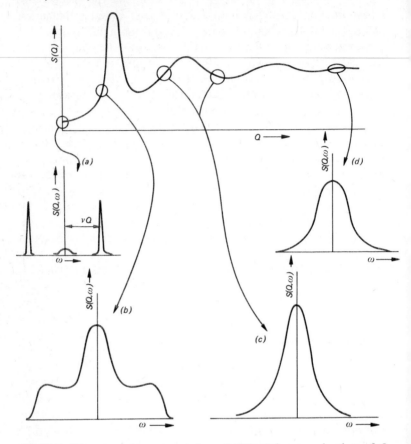

Fig. 7.5 Diagrammatic representation of $S(Q, \omega)$ for several values of Q. The upper curve is the function $S(Q)$, and the curves (a)–(d) show the spectral shape (variation of $S(Q)$ with ω) for the various fixed values of Q marked by circles on $S(Q)$ (after Egelstaff, 1967a).

Velocity autocorrelation function

At the optical limit, with zero momentum transfer it proves useful to evaluate the function

$$\hat{z}(\omega) = \omega^2 \lim_{Q \to 0} \left[\frac{S_s(Q, \omega)}{Q^2} \right] \tag{7.4}$$

This function $\hat{z}(\omega)$ is in fact the frequency distribution, "spectral density", or Fourier transform of the velocity autocorrelation function $z(t)$ (equation (5.9)) as shown in equation (A.11). Therefore if information on $\hat{z}(\omega)$ over a wide range of frequencies is available, neutron scattering permits the evaluation of $z(t)$.

In studying the experimentally determined velocity auto-correlation functions and spectral density plots, it is helpful to separate the short time, solid-like, damped vibratory or oscillatory behaviour from the long time behaviour which is diffusive and eventually becomes exponential decay. (For this purpose a "short" time for simple liquids is $t < 10^{-12}$ s.) An *oscillator* $z(t)$ has a transform $\hat{z}(\omega)$ which is a very sharp peak at the oscillator frequency. A damped oscillator has a broader peak in the frequency distribution. At the other extreme of long times, consider the simple Brownian diffusion model, for which the velocity autocorrelation function is

$$z(t) = \frac{kT}{m} \exp(-\gamma t) \qquad (7.5)$$

and which yields on Fourier transformation (see equation (A.7))

$$\hat{z}(\omega) = \frac{kT}{\pi m} \left(\frac{\gamma}{\omega^2 + \gamma^2} \right) \qquad (7.6)$$

In these equations, γ is the "friction constant" (compare equation (3.11))

$$\gamma = 1/\tau = \frac{kT}{mD} \qquad (7.7)$$

τ is the time spent in free diffusive movement; D is the diffusion coefficient; and m is the effective particle mass for diffusion, which includes the mass of the particle itself and the mass of neighbouring particles that have to be moved in order for diffusion to take place (Egelstaff, 1967a). This model is illustrated in Fig. 7.6, with the diffusive and damped vibratory components of the frequency distribution labelled I and II. The middle section of the figure shows how the diffusive period τ is described in terms of a vibratory part and a displacement part.

An example of the frequency distribution of the velocity auto-correlation function is shown in Fig. 7.7. This graph for water was obtained from neutron scattering ($0 < \beta < 5$) and Raman ($5 < \beta < 20$) data, and is plotted as a function of the dimensionless energy variable $\beta = \hbar\omega/kT$ (Egelstaff and Schofield, 1965). The peaks at large β values correspond to short time vibrations ($\beta = 17$ and 8) and rotations ($\beta = 2.5$). The peak at $\beta = 0.5$ is due to the liquid

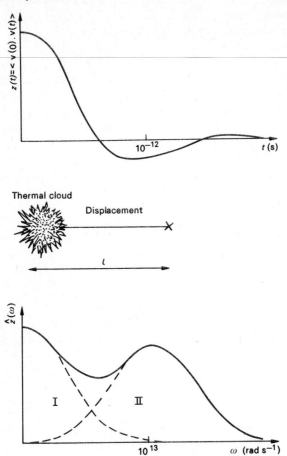

Fig. 7.6 The velocity autocorrelation function and its Fourier transform which is the frequency distribution $\hat{z}(\omega)$ with diffusive (I) and vibratory (II) components. In the middle section of the figure the diffusive period τ is described in terms of a vibratory and a displacement part of added length l (after Larsson, 1972).

state analogue of acoustic lattice vibrations (short-range order) and the peak at the origin (long times) corresponds to diffusive motions.

The macroscopic, low Q limit of radiation scattering

At low Q, $S(Q, \omega)$ exhibits sharp peaks at values of $\Delta\omega$ given by

$$\Delta\omega = \pm vQ \tag{7.8}$$

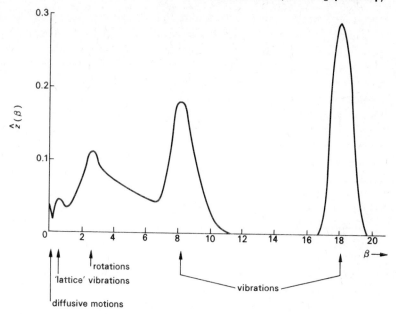

Fig. 7.7 Spectral density $\hat{z}(\beta)$ of the velocity autocorrelation function for water, in energy units of $kT(\beta = \hbar\omega/kT)$ (after Egelstaff and Schofield, 1965).

where v is the velocity of sound. For electromagnetic radiation, (from equation 7.2).

$$\Delta\omega = \pm 2\omega_0(v/c)\sin(\theta/2) \tag{7.9}$$

where ω_0 is the incident frequency and c is the velocity of light. These are known as the Brillouin doublet in optical spectroscopy; higher order peaks are also observed.

At this limit of $Q \to 0$, the concepts of hydrodynamics and thermodynamics may be used to discuss radiation scattering (Egelstaff and Schofield, 1965). The state of the liquid may be described in terms of the fluctuations of local values of the density, pressure, temperature and entropy. Pressure and entropy are convenient independent variables for these local properties: because it can be shown that their fluctuations are uncorrelated, they are statistically independent:

$$\Delta\rho = (\partial\rho/\partial p)_S\Delta p + (\partial\rho/\partial S)_p\Delta S \tag{7.10}$$

Entropy fluctuations cause local, non-propagating density fluctuations, and scatter radiation without change of frequency, yielding a central, undisplaced line, the Rayleigh line. The density fluctuations which arise from the pressure fluctuation produce periodic sonic disturbances or phonons which are propagated in all directions, but are observable by light scattering only when the so-called Bragg condition (analogous to the time-independent Bragg condition in X-ray crystal diffraction) applies. The sonic wave trains can be considered as a system of "moving diffraction gratings", so light from these "diffraction gratings" undergoes a Doppler shift as indicated in equations (7.8) and (7.9) and Fig. 7.5.

The pressure fluctuations are damped by *viscous* effects if the sound wave frequency is sufficiently high that there is appreciable rigidity or elasticity in the liquid, and the entropy or temperature fluctuations are damped by *thermal conduction* if Q is sufficiently large that the sound waves are no longer propagating adiabatically. At these intermediate values of Q the liquid may still be considered as a continuum, and hydrodynamic theory is adequate for a consideration of the effects of viscous and thermal conduction damping. As shown above, for still larger values of Q, of the order of the inverse interparticle separation, the detailed structure of the liquid must be considered.

Critical opalescence*

Usually simple fluids have no ordered structure for values of r which are considerably greater than intermolecular distances. However, when a fluid is near its critical point as discussed in Chapter 4 density fluctuations do extend over these distances, and result for visible radiation in opalescent light scattering. Analogous behaviour is observed for other forms of radiation, for example small angle X-ray diffraction which provides information on long interparticle distances (or macroscopic particles) (Brady, 1971).

One way of looking at the phenomena of critical density fluctuations and opalescence is that they are macroscopic versions of what occurs on microscopic distance and short-time scales in all liquids, and they therefore contribute to our understanding of the liquid state.

* Stanley, 1971.

General references

** BROOKHAVEN (1966), Proceedings of the International Conference on the Properties of Liquid Metals, "Structure and Scattering in Liquid Metals", *Adv. Phys.*, (1967), 16 (62, 63, 64), 147–307.

** CHEN, S. H. (1971), "Structure of Liquids", in *Physical Chemistry: An Advanced Treatise,* eds. Eyring, H., Henderson, D. and Jost, W., Vol. 8A, *Liquid State,* ed. Henderson, D., Academic Press, New York, Chap. 2, pp. 85–156.

*** CHU, B. and SCHMIDT, P. W. (1968), "Light Scattering from Simple Dense Fluids", in Frisch and Salsburg (1968), pp. 111–18.
A brief account of the limited results available for simple fluids.

*** EGELSTAFF, P. A. (1967a), *An Introduction to the Liquid State,* Academic Press, London, Chaps. 8, 9.

* EGELSTAFF, P. A. (1967b), "Radiation Scattering Studies of the Structure and Transport Properties of Liquids", *Disc. Faraday Soc.,* 43, 149–59.

* EGELSTAFF, P. A. and SCHOFIELD, P. (1965), "The Structure and Thermal Motion of Simple Liquids, Parts I and II", *Contemp. Phys.,* 6, 274–84, 453–64.
A very clear introductory account.

*** ENDERBY, J. E. (1968), "Neutron Scattering Studies of Liquids", in *Physics of Simple Liquids,* eds. Temperley, H. N. V., Rowlinson, J. S. and Rushbrooke, G. S., North-Holland, Amsterdam, Chap. 14, pp. 611–44.

** ENDERBY, J. E. (1972), "The Correlation Functions for Simple Liquids", *Adv. Struct. Res. Diffr. Meth.,* 4, 65–104.
A detailed account, with many experimental graphs, of the static structure factor and its theoretical significance.

*** FRISCH, H. L. and SALSBURG, Z. W. (1968), eds., *Simple Dense Fluids,* Academic Press, New York.
A collection of specialized chapters by different authors, with discussion and data on thermodynamic and transport properties, radiation scattering, dielectric data and nuclear and electronic spectroscopy of the simpler liquids.

** GINGRICH, N. S. (1965), "X-ray and Neutron Diffraction Studies of Liquid Structure", in *Liquids: Structure, Properties, Solid Interactions,* ed. Hughel, T. J., Proc. Symposium, Warren,

Michigan, 1963, Elsevier, Amsterdam, pp. 172–200.
A review of techniques and results for time-independent distribution functions.

★★★ INTERNATIONAL ATOMIC ENERGY AGENCY, Vienna (1961 on), *Series of Proceedings of Symposia on Inelastic Scattering of Neutrons:* 1st Vienna 1960, publ. 1961; 2nd, Chalk River, Canada, 1962, publ. 1963; 3rd, Bombay, 1964, publ. 1965; 4th, Copenhagen, 1968, publ. 1968.

★★ KRUH, R. F. (1962), "Diffraction Studies of the Structure of Liquids", *Chem. Rev., 62,* 319–46.
A detailed review of time-independent scattering data, mainly X-ray.

★★ LARSSON, K.-E. (1972), "Molecular Dynamics of Liquids in Relation to the Solid and Gas Phases as seen by Neutron Scattering", *Faraday Symp. Chem. Soc., 6,* 122–34.
N.B. The molecular dynamics technique is introduced in Chapter 12 of the present book.

★★★ LARSSON, K.-E., DAHLBERG, U. and SKÖLD, K. (1968), "Neutron Scattering Results", in Frisch and Salsburg (1968), pp. 119–82.
An introduction to the method and detailed results for helium, argon, hydrogen and methane.

★★★ MCINTYRE, D. and SENGERS, J. V. (1968), "Study of Fluids by Light Scattering", in *Physics of Simple Liquids,* eds. Temperley, H. N. V., Rowlinson, J. S. and Rushbrooke, G. S., North Holland, Amsterdam, Chap. 11, pp. 447–505.

★★★ MARSHALL, W. and LOVESEY, S. W. (1971), *Theory of Thermal Neutron Scattering,* Oxford University Press, London.
A comprehensive, detailed treatment.

★★ PALEVSKY, H. (1965), "Inelastic Neutron Scattering by Liquids", in *Liquids: Structure, Properties, Solid Interactions,* ed. Hughel, T. J., Proc. Symposium, Warren, Michigan, 1963, Elsevier, Amsterdam, pp. 201–17.

★★★ PINGS, C. J. (1968), "Structure of Simple Liquids by X-ray Diffraction", in *Physics of Simple Liquids,* eds. Temperley, H. N. V., Rowlinson, J. S. and Rushbrooke, G. S., North Holland, Amsterdam, Chap. 10, pp. 387–445.

** SCHMIDT, P. W. and TOMPSON, C. W. (1968), "X-ray Scattering Studies of Simple Fluids", in Frisch and Salsburg (1968), pp. 31–110.
Detailed results for liquids of the noble gases, homonuclear molecules and methane.

*** SPRINGER, T. (1972), "Quasielastic Neutron Scattering for the Investigation of Diffusive Motions in Solids and Liquids", *Springer Tracts in Modern Physics,* ed. Höhler, G., Vol. 64, Springer, Berlin–Heidelberg–New York.

Specific references

ALDRED, B. K., EDEN, R. C. and WHITE, J. W. (1967), "Neutron Scattering Spectroscopy of Liquids", *Disc. Faraday Soc.,* 43, 169–83.
A simple description of the technique, with some typical results.

ALDRED, B. K., STIRLING, G. C. and WHITE, J. W. (1972), "High Frequency Dynamics of Liquid Methanol and Toluene", *Faraday Symp. Chem. Soc.,* 6, 135–60.

BRADY, G. W. (1971), "Some Aspects of Small-Angle X-Ray Scattering", *Accounts Chem. Res.,* 4, 367–73.

DASANNACHARYA, B. A. and RAO, K. R. (1965), "Neutron Scattering from Liquid Argon", *Phys. Rev.,* 137, A417–27.

EGELSTAFF, P. A. (1965), ed., *Thermal Neutron Scattering,* Academic Press, London.
A specialized book emphasizing experimental aspects.

MIKOLAJ, G. and PINGS, C. N. J. (1967), "Structure of Liquids. III. An X-Ray Diffraction Study of Fluid Argon", *J. Chem. Phys.,* 46, 1401–11.

STANLEY, H. E. (1971), *Introduction to Phase Transitions and Critical Phenomena,* Clarendon Press, Oxford.

WINFIELD, D. J. and ROSS, D. K. (1972), "The quasi-elastic scattering of neutrons from C_6H_6 and C_6D_6", *Mol. Phys.,* 24, 753–72.

Chapter 8
Macroscopic electromagnetic liquid properties

Up to this point the electrical properties of liquids have been ignored except for a brief mention of dipole moment and polarizability. Electromagnetic phenomena in liquids form another example of macroscopic, experimentally measurable properties resulting from the interaction of microscopic or molecular properties. The macroscopic properties can be defined exactly, and in principle can be measured to any desired degree of precision, within the limitations of Heisenberg's uncertainty principle. The molecular properties on the other hand, although they may be defined exactly, have calculated numerical values which depend on the way in which we believe the microscopic properties are combined in the liquid, i.e. there is a variety of theories (Hill *et al.*, 1969; Glarum, 1972).

The relative permittivity, ε_r, probably more familiar as the "dielectric constant" defined as the ratio of the capacitances of material-filled to vacuum capacitors, may be expressed in terms of a set of vectors which satisfy electromagnetic equations of state known as Maxwell's equations. These *macroscopic* quantities are

electric field strength, **E**
electric displacement **D**
electric polarization **P**

and they are related by

$$\mathbf{D} = \varepsilon \mathbf{E} \tag{8.1}$$

$$\mathbf{P} = \mathbf{D} - \varepsilon_0 \mathbf{E} \tag{8.2}$$

where ε is the permittivity and ε_0 is the permittivity of a vacuum. The relative permittivity ε_r is defined

$$\varepsilon_r = \varepsilon/\varepsilon_0 \tag{8.3}$$

and may be called the dielectric constant when it is independent

74

of **E**. The refractive index n is related to ε_r to a good approximation in most non-polar liquids by

$$\varepsilon_r = n^2 \tag{8.4}$$

The subject of electric and magnetic properties appears confusing for the following reasons (McGlashan, 1971):

  (i) four systems of electric and magnetic *equations* are in use.

 (ii) in each case these equations may be written in both "rationalized" and "non-rationalized" forms. (In rationalized equations the factors 4π and 2π appear only where they are expected from the geometry, but in non-rationalized equations these appear when not expected from the geometry and do not appear when expected.)

(iii) Several systems of electric and magnetic *units* are in use.

The internationally recommended system of equations (which is used in this chapter) is based on four dimensionally independent quantities (length, mass, time and electric current) written in rationalized form. In this "four quantity" system of equations, the permittivity of a vacuum, ε_0, and the permeability of a vacuum, μ_0, appear as physical quantities which are not numbers. The relative permittivity $\varepsilon_r = \varepsilon/\varepsilon_0$ and relative permeability $\mu_r = \mu/\mu_0$ are, of course, dimensionless.

The polarization **P** of a dielectric material may be considered to be made up of the following contributions:

    electronic polarization $\Big\}$ making up the induced or
    atomic vibrational polarization $\int$ distortion polarization $\mathbf{P}_d$
    orientation polarization, $\mathbf{P}_0$, due to alignment of permanent dipoles.

Consider a parallel plate capacitor filled with a polar material, and with an alternating field applied. When the frequency of the applied field is sufficiently low, all the polarization processes can take place freely and reach the values they would have in a steady field equal to the alternating field at any instant. As the frequency is increased, a stage is reached where the polarization no longer has time to reach its steady value. The orientation polarization is the first to be affected, at frequencies of 10^{10}–10^{12} Hz (i.e. in the microwave and far infra-red regions) for small molecules, but at lower frequencies for larger molecules and at lower temperatures. At higher frequencies, comparable with the natural atomic vibration frequencies, the atomic polarization fails to attain its equilibrium value: the infra-red absorption region. The fall-off of the polarization caused by electronic transitions occurs at visible, ultra-violet and X-ray frequencies.

It is therefore apparent that a study of the frequency dependence of the absorption of electromagnetic radiation provides considerable information about liquid structure.

The bulk electromagnetic properties of a liquid depend on the molecular properties of polarizability α and permanent electric dipole moment **p**, but as might be expected there is no precise general relationship between molecular and macroscopic quantities for liquids. The permittivity of a dilute gas or dilute solution of polar material with a molecular concentration N/V is related to the electric field strength **E** by

$$(\varepsilon/\varepsilon_0 - 1)(V/N) = \alpha/\varepsilon_0 + p\langle\cos\theta\rangle/\varepsilon_0\mathbf{E} \qquad (8.5)$$

The first term on the right-hand side is due to the induced or distortion polarization, the molecular polarizability α being the proportionality factor between the magnitude of the *induced* dipole moment and the local electric field. The second term on the right is due to the orientation polarization, involving the *permanent* electric dipole moment of a molecule, p, and the average value of the angle between the dipole and the electric field. Other relationships have been developed more applicable to polar liquids which enable the dipole moment to be evaluated from the permittivities at static or low frequencies and at very high frequencies.

The electric dipole moment with dimensions (length) × (time) × (electric current) has as SI unit the metre-second-ampere or coulomb-metre. However, the unit called the debye, symbol D, from the non-rationalized electrostatic system is still widely used:

$$D \simeq 3.33564 \times 10^{-30} \text{ C m} \qquad (8.6)$$

It has been suggested (McGlashan, 1971) that the quantity

$$l_p = p/e \qquad (8.7)$$

(where e denotes the charge of a proton) which has dimensions of length (the "dipole length") may be a more convenient quantity than p, and this is included in the physical properties collected in Table 14.1.

Once again in electromagnetic properties it is apparent that dynamic, spectroscopic measurements provide more information on details of molecular motion than the static or time-independent properties do, and electromagnetic spectroscopy is the subject of the next chapter.

General references

*** AMEY, R. L. (1968), "The Electromagnetic Equation of State Data", in *Simple Dense Fluids,* eds. Frisch, H. L. and Salsburg, Z. W., Academic Press, New York, pp. 183–201.
A brief review of the electromagnetic theory and data for simple liquids.

 * CROSSLEY, J. (1971), "Dielectric Relaxation and Molecular Structure in Liquid", *R.I.C. Rev.,* 4, 69–96.

*** DEBYE, P. (1929), *Polar Molecules,* Reinhold; reprinted (1945) Dover Publ. Co., New York.
A classic work on the subject.

 ** HILL, N. E., VAUGHAN, W. E., PRICE, A. H. and DAVIES, M. (1969), *Dielectric Properties and Molecular Behaviour,* Van Nostrand Reinhold, London.
A comprehensive account of many aspects of the electrical behaviour of molecular systems.

Specific references

GLARUM, S. H. (1972), "Conflicting Theories of Dielectric Relaxation", *Mol. Phys.,* 24, 1327–39.

MCGLASHAN, M. L. (1971), *Physico-Chemical Quantities and Units,* Royal Institute of Chemistry Monographs for Teachers No. 15, Chapter 12.

Chapter 9
Electromagnetic spectroscopy and dipole correlation functions

Molecules in the gas phase undergo free rotation between collisions, and information about the rotational energy levels is contained in the rotational fine structure. In liquids the vibrational bands are broadened, and resolvable fine structure is lost, but information about the nature of rotational and translational motion in the liquid is still contained in the broadened infra-red vibration band.

When light of a frequency corresponding to an allowed transition between quantum states illuminates a molecule, the probability that a transition takes place is a function of the electric field, the dipole moment, and the probability of finding molecules in the initial state. It can be shown that the shape of the absorption band $I(\omega)$ is related by a Fourier transformation to the dipole moment autocorrelation function. An illustration of the dipole autocorrelation function of CO in different environments is shown in Fig. 9.1. In physical terms, this describes how a vibrating dipole oriented in a certain direction in the liquid changes its direction with time. During a brief initial period, free rotation determines the kinetics of rotation. This is followed by a libration ("rotational oscillation") involving a change in sign of the dipole autocorrelation function, and finally reorientation becomes essentially random as the autocorrelation function tends to zero.

Dielectric relaxation and far infra-red absorption

Radiofrequency and microwave dielectric relaxation and far infra-red absorption spectroscopy of dipolar liquids probe the correlation function of the total electrical polarization of the system. Cross terms $\mathbf{p}_i(0) \cdot \mathbf{p}_j(t)$ in the dipole correlation function are important, and the absorption results cannot be simply interpreted in terms of the orientation of a single molecule. As can be seen from reference to Table (6.1) this corresponds to dipole or collisionally induced dipole rotation which tends to be "free" or uncorrelated at dielectric

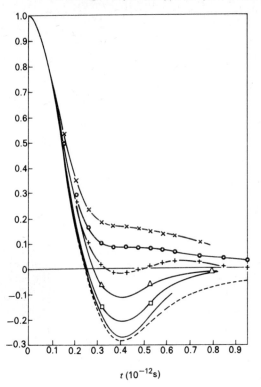

Fig. 9.1 Dipole autocorrelation function of CO in different environments: ×, CO in liquid $CHCl_3$; ○, CO in liquid CCl_4; +, CO in liquid n-C_7H_{16}; △, □, —, CO in argon gas at decreasing pressures; ---, CO free (calculated) (after Gordon, 1965).

relaxation frequencies, and a correlated librational movement in the far infra-red. Many theoretical approaches have been made to the explanation of dielectric relaxation and far infra-red absorption spectra (Hill *et al.*, 1969; Crossley, 1971), but it appears that the dipole correlation formalism (Williams, 1972; van Aalst *et al.*, 1972) is best suited to discuss the results. Experimentally the frequency dependence of the absorption shows a broad, featureless band, with bandwidths equal to or even larger than the frequency of maximum absorption (Fig. 9.2).

Vibration spectroscopy in the near infra-red

The interpretation of the dipole correlation function is often simpler in near infra-red spectroscopy than in dielectric and far infra-red

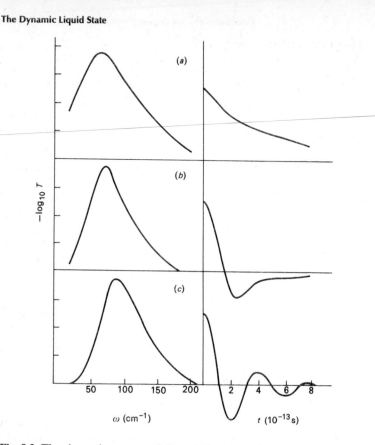

Fig. 9.2 The absorption spectra (left) and dipole correlation functions (right) of: (a) CH$_3$CN at 314 K; (b), HCl in liquid argon at 105 K and 5.7 atm; and (c), molten caesium nitrate at 653 K (after van Aalst *et al.*, 1972).

spectroscopy. This is because it is often a good approximation to consider the internal vibrational motion of the molecules *separable* from the rotational and translational motion. This means that a vibrational excitation is localized on a single molecule and the correlation function is that of the vibrational dipole moment term of a typical molecule, without contributions from cross terms. The Fourier transformation of a near infra-red frequency spectrum then yields the dipole correlation function shown above in Fig. 9.1. The negative correlation in gaseous systems indicates that after a time it is more probable that a molecule has swung around to point in the opposite direction from that which it had at $t = 0$. This reorientation is not permitted as readily in a liquid.

Raman light scattering

For radiation whose frequency is well away from absorption frequencies of a medium, scattering rather than absorption is the most probable process. Raman scattering may be described in quantum mechanical terms as an inelastic collision between a photon and a group of interacting molecules, or in classical terms as follows (Tobias, 1967). The oscillating electric field in the radiation induces in a molecule an electric dipole moment the magnitude of which depends on the amplitude of the light wave and the polarizability of the molecule. The field fluctuates at the frequency of the light, ω_0, and the polarizability varies because of the frequency ω_1 of the molecular vibration or rotation. The resulting expression for the oscillation of the induced dipole moment contains terms in the exciting frequency ω_0, and also terms in the "beat" frequencies of the light and molecular vibrational or orientational frequencies $(\omega_0 + \omega_1)$ and $(\omega_0 - \omega_1)$. Oscillation of the dipole is accompanied by radiation at these frequencies.

The extent of polarization of the Raman scattering provides additional information. If unpolarized light is used to excite the molecules, the scattering (observed at 90° to the incident light) is at least partially polarized. Experimentally, polarized light is used to excite the spectra, and the scattered radiation intensity can then be separated into polarized and depolarized components. It can be shown that the polarized intensity is determined by the Fourier transform of the polarizability correlation function $\langle \alpha(0)\alpha(t) \rangle$. The depolarized scattering is determined by the correlation of the anisotropy of the polarizability, and if the molecule has some symmetry (a threefold or higher axis) this correlation function has the form $\langle \frac{1}{2}[3(\mathbf{p}(0) \cdot \mathbf{p}(t))^2 - 1] \rangle$.

It can be seen therefore that although dielectric relaxation, infra-red absorption and Raman scattering are all determined by the orientation and vibration of dipolar or induced dipolar molecules, each technique provides information on a particular correlation function.

The various types of Raman scattering are included in Table 6.1. When the scattering process excites a molecular vibration, the *vibrational Raman* effect is observed, with polarized (isotropic) and depolarized (anisotropic) components. If there is no change in the vibrational term during scattering, the polarized term corresponds to *Rayleigh scattering* and the depolarized scattering is the *rotational Raman* effect. The Rayleigh scattering has additional components (Brillouin scattering) caused by the Doppler effect which has already been discussed. In correlation function formalism, Brillouin scattering may be described in terms of a correlation function which includes correlations in space as well as time.

Extremely good resolving power is necessary to determine the shapes of the very narrow Rayleigh and Brillouin peaks, and conventional spectrometers are inadequate for this purpose. However, the new techniques of optical mixing spectroscopy, both optical heterodyne and homodyne or self-beating (Pike, 1970; Stanley, 1971), operate on the principle of translating the spectral information from a spectrum centred on the optical frequency ($\omega \approx 10^{15}$ Hz) to a spectrum centred about $\omega = 0$, and thus permit it to be more easily resolved.

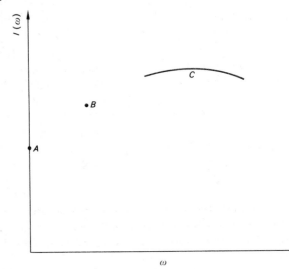

Fig. 9.3 Diagrammatic representation of the information about a frequency-dependent property $I(\omega)$ provided by various techniques: A, at limiting low frequency from a static measurement; B, at a single finite frequency from n.m.r.; C, over a range of frequencies from absorption or scattering spectroscopic techniques.

Nuclear magnetic resonance*

If nuclear magnetic spins are coupled weakly to the molecular "lattice" motion, the correlation functions of the spin variables observed experimentally decay much more slowly than the correlation functions of the molecular motions in which one is interested. In this case the spin correlations may be

---

* Deutch and Oppenheim, 1968.

simply expressed as an exponential function of time, proportional to $\exp(-t/T)$ with a time constant T. If a uniform magnetic field is applied, then even for an isotropic liquid the system has only cylindrical symmetry, and it is necessary to distinguish the decay of the longitudinal correlation (with a longitudinal relaxation time T_1) and the transverse correlation (with a transverse relaxation time T_2). For a spin of angular momentum $S = \frac{1}{2}$, the only interactions are magnetic. The spectral density (Fourier transform) $I(\omega)$ of the correlation function of the local magnetic field caused by the molecular motion is given by equation (A.14), and the relaxation times T_1 and T_2 are related to this spectral density at one particular frequency, the Larmor precession frequency of the molecule in the magnetic field. Thus it can be seen that measurement of "static" properties such as diffusion coefficients gives information at the *zero limit* of frequency in the spectral density of a molecular correlation function; absorption and scattering spectroscopy techniques provide data on a *range* of frequencies; but the nuclear magnetic resonance technique looks at a particular finite frequency (Fig. 9.3).

For a spin $S = 1$ there is the possibility of electric quadrupole interactions in addition to the magnetic interactions, but for the simple cases where the spin lies on a threefold or higher symmetry axis, the spectral density is the Fourier transform of the correlation function $\langle \frac{1}{2}[3(\mathbf{p}(o) \cdot \mathbf{p}(t))^2 - 1] \rangle$, which is precisely the same correlation function which determines the shape of the depolarized part of a Raman band in a molecule of similar symmetry.

General references

★ CROSSLEY, J. (1971), "Dielectric Relaxation and Molecular Structure in Liquids", *R.I.C. Rev.*, 4, 69–96.

★★ DEUTCH, J. M. and OPPENHEIM, I. (1968), "Time Correlation Functions in Nuclear Magnetic Resonance", in *Advances in Magnetic Resonance,* ed. Waugh, J. S., Vol. 3, Academic Press, New York, pp. 43–78.

★★ GORDON, R. G. (1968), "Correlation Functions for Molecular Motion", in *Advances in Magnetic Resonance,* ed. Waugh, J. S., Academic Press, Vol. 3, pp. 1–42.

★★ HILL, N. E., VAUGHAN, W. E., PRICE, A. H. and DAVIES, M. (1969), *Dielectric Properties and Molecular Behaviour,* Van Nostrand Reinhold, London.

★★ ROBIN, M. B. (1968), "Spectroscopy in Simple Liquids", in *Simple Dense Fluids,* eds. Frisch, H. L. and Salsburg, Z. W., Academic Press, New York, pp. 215–50.
An account of the results obtained for the rare gases, H_2, CO, O_2, N_2 and CH_4.

★ TOBIAS, R. S. (1967), "Raman Spectroscopy in Inorganic Chemistry I. Theory", *J. Chem. Educ.*, 44, 1–8.

★★ VAN AALST, R. M., VAN DER ELSKEN, J., FRENKEL, D. and WEGDAM, G. H. (1972), "Interpretation of Dipole Correlation Functions in Some Liquid Systems", *Faraday Symp. Chem. Soc.*, 6, 44–105.

★★★ WILLIAMS, G. (1972), "The Use of the Dipole Correlation Function in Dielectric Relaxation", *Chem. Rev.*, 22, 55–69.

★★ YOUNG, R. P. and JONES, R. N. (1971), "The Shapes of Infrared Absorption Bonds of Liquids", *Chem. Rev.*, 71, 219–28.

Specific references

GORDON, R. G. (1965), "Molecular Motion in Infrared and Raman Spectra", *J. Chem. Phys.*, 43, 1307–12.

PIKE, E. R. (1970), "Optical Spectroscopy in the Frequency Range $1–10^8$ Hz", *Rev. Physics in Technology*, 1, 180–4.

STANLEY, H. E. (1971), *Introduction to Phase Transitions and Critical Phenomena*, Clarendon, Oxford, Chap. 14, "Measurement of the Dynamic Structure Factor for Fluid Systems".

Chapter 10
Models of the liquid state

A simple physical model of a liquid, constructed of ball bearings for example, is an attempt to provide an instantaneous picture of a liquid on a molecular scale. The following is a summary of the attempts at this kind of experiment which were prompted by the fact that all experiments on real liquids yield only bulk or averaged properties.

(i) Heaps of deformable spheres★

After heaps of spheres made from materials such as wax, "Plasticene" or gelatine have been compressed, the balls are transformed into a set of polyhedra, the shapes of which provide information on the way in which the spheres were arranged. It is noticeable

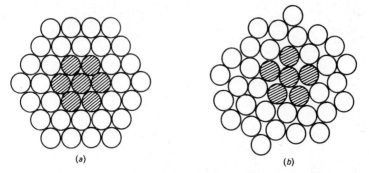

(a)   (b)

Fig. 10.1 Random close packing with five-fold symmetry (*b*) compared with regular or crystal close-packing (*a*) in two dimensions (after Smith, 1971).

that pentagonal faces predominated, i.e. in any plane through a reference molecule there is a tendency to a co-ordination number of five in an irregular dense assembly. This situation should be contrasted with the two-dimensional co-ordination number of six in a regular close-packed arrangement of spheres (Fig. 10.1). Bernal believed that irregular dense packing and pentagonal arrangements were necessarily connected.

★ Bernal, 1959; Hildebrand *et al.,* 1970.

(ii) Assemblies of hard spheres

Typically several thousand ball bearings are poured into a balloon with a specially prepared irregular surface (to prevent ordered layers being formed). They are shaken down, and compressed with rubber bands. In one experiment (Bernal and Mason, 1960) paint was poured into the random close-packed assembly of spheres and allowed to drain away. The paint remained where the spheres were close, and the spheres had on their surfaces a dot where another sphere had nearly touched and a ring where the spheres had exactly touched (Fig. 10.2). This provided statistical information on the co-ordination of randomly close packed spheres (Bernal, 1964; Bernal and King, 1967; Bernal and Finney, 1967; Finney, 1970; Caron, 1971). The observation (Bernal, 1959, 1960, 1965) that for ball bearing assemblies there is a reproducible maximum limiting state of *random* close packing using 0.634 of available space in

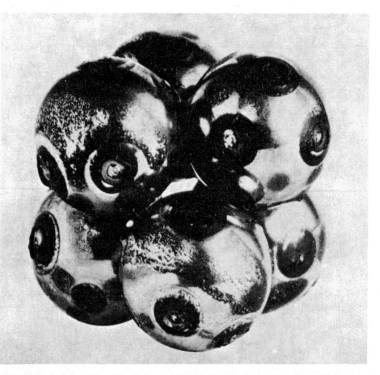

Fig. 10.2 Portion of steel sphere assembly showing contacts and near contacts (from Bernal and King, 1968, with the permission of the Royal Society, London).

addition to the hexagonal close packed or face-centred cubic *ordered* packings which have a "density" of 0.7405, i.e. $(\pi\sqrt{2}/6)$, seems fundamental to an understanding of liquid structure. Liquids may be compressed continuously to the limiting random or disordered close-packed density, but any transition to the ordered close-packed density is discontinuous because there is no homogeneous state with an intermediate density. Liquids are qualitatively different from crystalline solids, and the difference is a matter of geometry. Everyday experience in stacking cylindrical tins on a shelf provides a simple example of the fundamental difference between random and ordered close packing. Packing densities of mixtures of hard spheres of different diameters have also been studied (Dexter and Tanner, 1971).

(iii) Ball and spoke models

Methods (i) and (ii) attempt to duplicate the limiting high density hard sphere liquid state. As an alternative, ball and spoke models have been made, with spokes added randomly, or built up to display the results of random steel ball experiments (Fig. 10.3). It was found that the most commonly occurring configuration was the tetrahedron, and although regular tetrahedra cannot be fitted together to fill space, they can extend in a one-dimensional direction like a helix. It is interesting that there were in the models almost straight lines of particles, up to six in length, oriented randomly.

(iv) Two-dimensional "dynamic" models

A two-dimensional "dynamic" liquid simulator has been devised (Walton and Woodruff, 1969) for monatomic liquids in which atomic cohesion and repulsion are simulated by oil-covered ball bearings, and "thermal energy" is provided by the vibration of rough surfaced glass. This simulator permits the study of equilibrium (pair distribution functions), transport (viscosity and diffusion) and melting properties. Figure 10.4 shows a close-up view of the simulated liquid.

Zarzycki (1969) has simulated two-dimensional ionic liquids by means of floaters containing magnets, and has computed van Hove correlation functions. Thermal motion was simulated by waves on the surface of the water, and the "dynamic liquid structure" was recorded with a movie camera.

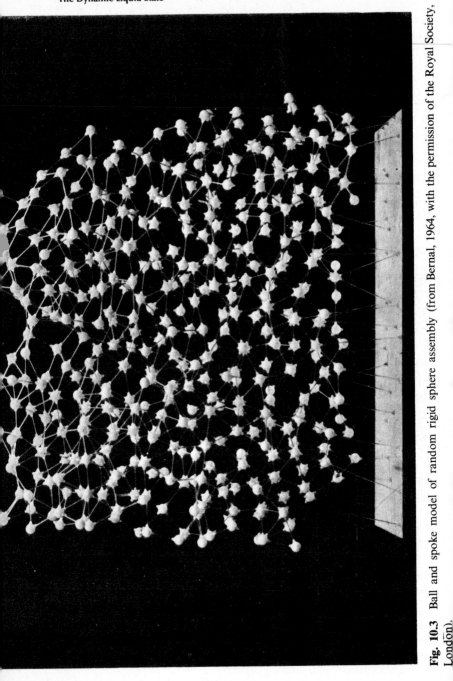

Fig. 10.3 Ball and spoke model of random rigid sphere assembly (from Bernal, 1964, with the permission of the Royal Society, London).

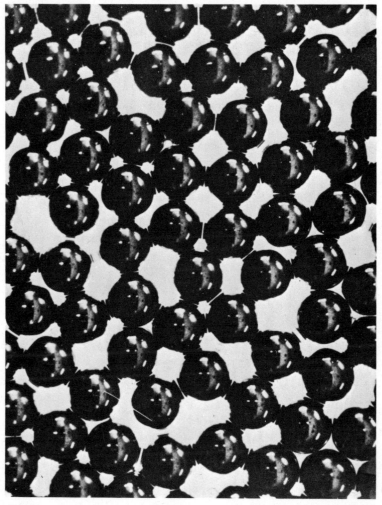

Fig. 10.4 A typical array of "molecules" in a two-dimensional liquid simulator (from Walton and Woodruff, 1969).

(v) Computer simulation of random packing of spheres *

The problem of random packing of spheres has been approached recently by means of a computer simulation of the physical process of dropping spheres into a bin: two- and three-dimensional arrays with a unidirectional gravitational force. (Computer simulation is discussed further in Chapter 12.)

* Visscher and Bolsterli, 1972.

General references

★★ BERNAL, J. D. (1964), "The Structure of Liquids", *Proc. Roy. Soc. (London)*, A280, 299–322.

★★ BERNAL, J. D. and KING, S. V. (1968), "Experimental Studies of a Simple Liquid Model", in *Physics of Simple Liquids,* eds. Temperley, H. N. V., Rowlinson, J. S. and Rushbrooke, G. S., North-Holland, Amsterdam, Chap. o, pp. 231–52.

★ DREISBACH, D. (1966), *Liquids and Solutions,* Houghton Mifflin, Boston, U.S.A., Chap. 6.

★★ HILDEBRAND, J. H., PRAUSNITZ, J. M. and SCOTT, R. L. (1970), *Regular and Related Solutions,* Van Nostrand Reinhold, New York, Chap. 3.

★ PRYDE, J. A. (1966), *The Liquid State,* Hutchinson, London, pp. 136–41.

Specific references

BERNAL, J. D. (1959), "A Geometrical Approach to the Structure of Liquids", *Nature,* 183, 141–7.

BERNAL, J. D. (1960), "Geometry of the Structure of Monatomic Liquids", *Nature,* 185, 68–70.

BERNAL, J. D. (1965), "The Geometry of the Structure of Liquids", in *Liquids: Structure, Properties, Solid Interactions,* ed. Hughel, T. J., Proc. Symposium, Warren, Michigan, 1963, Elsevier, Amsterdam, pp. 25–50.

BERNAL, J. D. and FINNEY, J. L. (1967), "Random Close-Packed Hard Sphere Model. II. Geometry of Random Packing of Hard Spheres", *Disc. Faraday Soc.,* 43, 62–9.

BERNAL, J. D. and MASON, J. (1960), "Co-ordination of Randomly Packed Spheres", *Nature,* 188, 910–1.

BERNAL, J. D. and KING, S. V. (1967), "Random Close-Packed Hard Sphere Model. I. Effect of Introducing Holes", *Disc. Faraday Soc.,* 43, 62–9.

CARON, L. G. (1971), "Bernal Model: A Simple Equilibrium Theory of Close-Packed Liquids", *J. Chem. Phys.,* 55, 5227–32.

DEXTER, A. R. and TANNER, D. W. (1971), "Packing Density of Ternary Mixture of Spheres", *Nature Phys. Sci.,* 230, 177–9.

FINNEY, J. L. (1970), "Random Packings and the Structure of Simple Liquids. I. The Geometry of Random Close Packing", *Proc. Roy. Soc. (London),* A319, 479–93.

SMITH, B. L. (1971), *The Inert Gases: Model Systems for Science,* Wykeham Publications, London, Chap. 6.

WALTON, A. J. and WOODRUFF, A. G. (1969), "A Kinetic Liquid Simulator", *Contemp. Phys.,* 10, 59–70.

VISSCHER, W. M. and BOLSTERLI, M. (1972), "Random Packing of Equal and Unequal Spheres in Two and Three Dimensions", *Nature,* 239, 504–7.

ZARZYCKI, J. (1969), "Un Modèle Analogique Dynamique d'un Liquide Ionique", *J. Chim. phys.,* Special no. Oct. 1969, 153–66.

Chapter 11
Effective pair potential energy functions and the pairwise additivity assumption

For at least 100 years (van der Waals, 1873) physical chemists and physicists have been attempting to explain in detail the macroscopic, experimentally measurable properties of liquids in terms of the individual, microscopic or molecular properties and intermolecular forces. All these various attempts are theories of the liquid state and constitute one aspect of the general atomic and molecular theory of matter which has proved so successful in explaining the structure of molecules and describing the intermolecular properties of the solid and gaseous states. For a long time the experimental quantities available were restricted to static or time-independent properties: thermodynamic quantities and the transport coefficients of diffusion, viscosity and, where appropriate, electrical conductivity; but recently spectroscopic dynamic or time-dependent data have become available also.

It is necessary to develop methods for calculating values of these macroscopic properties for comparison with experimental values, and as usual when applying the laws of mechanics to molecules, statistical methods must be used. For most liquids (those other than the light liquids, notably helium) the molecular interactions may be treated by classical mechanics. The resulting techniques are statistical thermodynamics to evaluate thermodynamic properties, and non-equilibrium or irreversible statistical thermodynamics to evaluate transport properties.

Two concepts have proved of great importance in this work: the effective pair potential energy function, and the pairwise additivity assumption.

Attractive and repulsive forces

In Chapter 2 the idea of a separation of intermolecular forces into attractive and repulsive components was introduced. The *attractive* or binding components of intermolecular forces are

 (i) valence
 (ii) ionic or coulombic
(iii) metallic
(iv) van der Waals (electrostatic and dispersion or London).

Valence and ionic bonds are very strong: materials with these types of interaction are generally solids under ambient conditions. Metallic forces depend on the presence of free electrons, and metals differ widely in their melting points (compare mercury and tungsten, for example). Van der Waals binding forces are weaker, and in general are responsible for the cohesion of those materials which are liquids at or below ambient temperature. Consequently these forces are of great interest in any consideration of the liquid state and they are discussed further below. *Repulsive* forces arise between all molecules when full shells of electrons are brought close together. In addition there are repulsive interactions between charged and polar species.

The effective pair potential energy function

In order to describe intermolecular forces quantitatively, the concept of an effective intermolecular potential energy function has proved valuable. The (hypothetical) pair potential energy of two isolated molecules as a function of intermolecular distance is expressed as the sum of two or more terms, each representing a contributing repulsive or attractive force, i.e. a positive or negative energy contribution. Even though it may never be possible to predict or describe *exactly* the potential energy of a pair of isolated molecules at all separations, the *effective* pair potential energy function can be established and described to any reasonable required precision by a series of terms with adjustable parameters. Experimental methods to determine the nature of the pair potential energy function include thermodynamic and transport properties of gases, equilibrium properties of condensed phases, and molecular beam experiments (Scott, 1971). This subject will not be discussed here. The task of describing pair potential energy functions is simplified if the terms are approximately correct, and fortunately the basic contributions may be expressed as the sum of several rather simple terms. It is convenient in discussing these potential energy functions to classify them as long range or short range.

Long-range potential energy functions

These may be expressed as a power series of inverse intermolecular distances. For two charged particles (in an ionic liquid, for example) the classical electrostatic or coulombic potential energy function is

$$\phi_c(r) \; \alpha \; \frac{1}{r} \tag{11.1}$$

and for the electrostatic energy arising from interaction of permanent dipoles, the first term of the effective pair potential energy expression is

$$\phi_s(r) \; \alpha \; (r^6 kT)^{-1} \tag{11.2}$$

The factor (kT) indicates that this is not a true intermolecular property and emphasizes the *effective* nature of the pair potential energy function. Even if a molecule does not have a permanent dipole moment, an electrical field acting on it induces a dipole moment with a resulting energy $\phi_i(r)$ from interaction of induced dipoles. Also, even in the case of uncharged atoms with a spherically symmetric charge distribution there is a mutual attraction due to instantaneously induced dipoles: the dispersion or London term $\phi_d(r)$ which is the main contributor to the van der Waals forces in non-polar molecules. To a reasonable approximation the three contributions to the total interaction can be treated separately and additively:

$$\phi(r) = \phi_s(r) + \phi_i(r) + \phi_d(r) \tag{11.3}$$

The induction and dispersion terms also have as their first term an inverse 6th power dependence on the intermolecular distance (equation (11.2)). This result follows both from classical considerations and also from quantum mechanical calculations.

Short-range potential energy functions

When the distance between molecules is small enough for their electron clouds to overlap, a quantum mechanical treatment must be used. One possibility is that chemical bonding will occur, but if this does not happen the alternative is that a potential corresponding to an extremely strong repulsive force exists due to the operation of the Pauli exclusion principle. Usually this potential energy function is represented as an exponential expression:

$$\phi(r) \; \alpha \; \exp(-cr) \tag{11.4}$$

or as a high inverse term in intermolecular distance:

$$\phi(r) \propto r^{-n} \tag{11.5}$$

with n usually 12 or greater.

The properties of assemblies of molecules interacting with varying degrees of "softness" (4th, 6th, 9th, 12th power and hard sphere) have been reviewed by Hoover *et al.*, 1971.

Examples of effective pair potential energy functions

A variety of empirical expressions has been developed as effective pair potential energies, with a range of compromises between mathematical convenience on the one hand and physical reality on the other. These are summarized below in approximate order of decreasing simplicity and increasing reality.

(a) Hard sphere without attraction

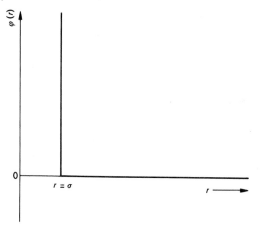

Fig. 11.1 The hard sphere function (equation (11.6)).

The hard sphere model (Fig. 11.1) may be described

$$\phi(r) = \begin{cases} +\infty \text{ for } r < \sigma \\ 0 \text{ for } r \geqslant \sigma \end{cases} \tag{11.6}$$

This implies an infinitely steep repulsive force, and no attractive force.

95

(b) Point centres of attraction or repulsion

$$\phi(r) = cr^{-n} \tag{11.7}$$

If c is positive and n is very large this approaches the hard sphere potential in character. It is also used for the coulombic potential between two ions with $n = 1$ and c positive for ions of like charge, negative for unlike charge.

(c) van der Waals function: hard sphere with
    uniform attractive potential energy

It can be shown (Kac, 1959; Kac et al., 1963; Rowlinson, 1970) that the van der Waals equation (2.6) corresponds to the pair potential energy function of a pair of hard spheres surrounded by an energy well whose depth and slope tend to zero and whose range tends to infinity:

$$\phi(r) = \begin{cases} +\infty \text{ for } r < \sigma \\ \lim_{\gamma \to \infty} \dfrac{-\alpha}{\gamma^3} \exp \dfrac{\sigma^3 - r^3}{\gamma^3} \text{ for } r \geqslant \sigma \end{cases} \tag{11.8}$$

where α is a positive constant.

(d) Square well: hard sphere with limited attraction region

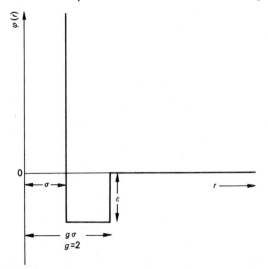

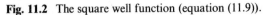
Fig. 11.2 The square well function (equation (11.9)).

This potential, illustrated in Fig. 11.2, is described mathematically:

$$\phi(r) = \begin{cases} +\infty \text{ for } r < \sigma \\ -\varepsilon \text{ for } \sigma \leqslant r \leqslant g\sigma \\ 0 \text{ for } r > g\sigma \end{cases} \tag{11.9}$$

where frequently $g = 2$.

(e) Triangular well

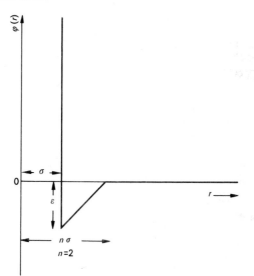

Fig. 11.3 The triangular well function (equation (11.10)).

Figure 11.3 illustrates the potential

$$\phi(r) = \begin{cases} +\infty \text{ for } r < \sigma \\ [n\varepsilon/(n-1)]\,[r/\sigma n - 1] \text{ for } \sigma \leqslant r \leqslant n\sigma \\ 0 \text{ for } r > n\sigma \end{cases} \tag{11.10}$$

(f) Sutherland function

This is a combination of (a), hard sphere, and (b), point centre of attraction:

$$\phi(r) = \begin{cases} \infty \text{ for } r < \sigma \\ -cr^{-n} \text{ for } r \geqslant \sigma \end{cases} \tag{11.11}$$

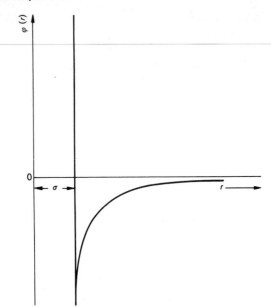

Fig. 11.4 The Sutherland $(\infty, 6)$ function (equation (11.11)).

For $n = 6$, corresponding to the predicted inverse 6th attractive term illustrated in Fig. 11.4, this is known as the Sutherland $(\infty, 6)$ function.

(g) Lennard-Jones (12,6) function

A closer approach to reality is made by the effective pair potential energy function shown in Fig. 11.5:

$$\phi(r) = 4\varepsilon\left[\left(\frac{\sigma}{r}\right)^{12} - \left(\frac{\sigma}{r}\right)^{6}\right] = \left[\left(\frac{r_o}{r}\right)^{12} - 2\left(\frac{r_o}{r}\right)^{6}\right] \qquad (11.12)$$

ε is the minimum potential energy, corresponding to $r = r_o$. This function is reasonably satisfactory for the *liquid* noble gases but not for the dilute gases, and has been widely used although it is now clear that the true noble gas pair potential energy function is both deeper and steeper than the Lennard–Jones (12, 6).

Angular dependence

In this discussion the angular dependence of interactions between molecules has been neglected, i.e. only interactions between *atoms*

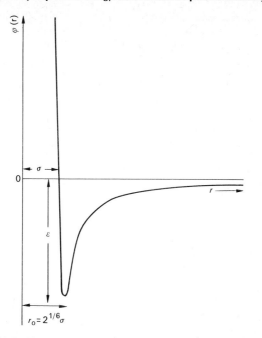

Fig. 11.5 The Lennard–Jones (12, 6) function (equation (11.12)).

have been considered. Although this complication has considerable significance for polyatomic systems, only one polyatomic case will be considered: that of nearly spherical or quasi-spherical molecules where the angular dependence is small and may be neglected.

Quasi-spherical molecules

In polyatomic molecules the origins of the attractive and repulsive forces are not the geometrical centres of the molecules, but interactions occur between the electrons of the peripheral atoms. Consequently the attractive and repulsive forces decrease with increasing separation more rapidly than would be expected according to the Lennard–Jones (12, 6) function, for example, and cause an increase in the apparent exponents of the function when these are referred to the centres of the molecules. In the case of the "quasi-spherical" or "globular" molecules such as SF_6, CF_4 and $C(CH_3)_4$ the molecules can be assumed to be approximately spherical (Hamann and Lambert, 1954).

Equation (11.12) is in fact a special case of the Lennard–Jones bireciprocal or (n, m) function which may be expressed for $n > m > 3$:

$$\phi(r) = \frac{\varepsilon}{n-m}\left[m\left(\frac{r_o}{r}\right)^n - n\left(\frac{r_o}{r}\right)^m\right] \tag{11.13}$$

99

For quasi-spherical molecules the potential energy function can be expressed to a good approximation by equation (11.13) with $n \approx 30$ and $m = 7$. A value of $n = 28$ is often used for mathematical convenience:

$$\phi(r) = \frac{\varepsilon}{3}\left[\left(\frac{r_o}{r}\right)^{28} - 4\left(\frac{r_o}{r}\right)^{7}\right] \qquad (11.14)$$

The "attractive", negative part is slightly steeper, and the "repulsive", positive part is considerably steeper than the corresponding parts of the (12, 6) potential.

An alternative approach was taken by Kihara (1953) who assumed that the Lennard–Jones potential function existed between hard cores inside the molecules:

$$\phi(r) = \frac{\varepsilon}{n-m}\left[m\left(\frac{\rho_o}{\rho}\right)^{n} - n\left(\frac{\rho_o}{\rho}\right)^{m}\right] \qquad (11.15)$$

where ρ is the shortest distance between cores, and ρ_0 is its value at the potential energy minimum. This Kihara (12, 6) function has the same effect as the Lennard–Jones (28, 7) function, steepening the positive or repulsive more than the negative (attractive) contribution. An example of the application of various potential energy functions to quasi-spherical molecules is the paper by Dawe *et al.* (1970).

The pairwise additivity assumption

For simplicity it has been usual to assume that the potential energy of a fluid is the sum of the potential energies of every pair of molecules each pair being treated as if isolated. For a collection of fixed electric charges, classical electrostatics shows that the potential energy is a sum over pairs, and hard sphere positive energies would also be expected to be additive. However, for polarizable electron clouds this only approximates to the true situation (Scott, 1971; Enderby, 1972).

One definition of the commonly used term *simple liquid* is a liquid made up of molecules which interact in this simplified, idealized, pairwise additive manner.

General references

** BUCKINGHAM, A. D. (1970), "Intermolecular Forces", *Pure Applied Chem.*, 24, 123–34.

 * DREISBACH, D. (1966), *Liquids and Solutions*, Houghton Mifflin, Boston, U.S.A., Chap. 9.

** EGELSTAFF, P. A. (1967), *An Introduction to the Liquid State,* Academic Press, London, Chaps. 2–4.

** ENDERBY, J. E. (1972), "The Correlation Functions for Simple Liquids", *Adv. Struct. Res. Diffr. Meth.,* 4, 65–104.

*** HIRSCHFELDER, J. O., CURTISS, C. F. and BIRD, R. B. (1954), *Molecular Theory of Gases and Liquids,* Wiley, New York.
A "standard work" on gases and liquids from a molecular and statistical thermodynamic viewpoint.

*** MARGENAU, H. and KESTNER, N. R. (1969), *Theory of Inter-Molecular Forces,* Pergamon, Oxford.
This includes developments in the theory of intermolecular forces since the book by Hirschfelder et al. (1954).

** PRYDE, J. A. (1966), *The Liquid State,* Hutchinson, London, Chap. 4.

** ROWLINSON, J. S. (1965), "The Equation of State of Dense Systems", *Rpts Progr. Phys.,* 28, 169–99.

** ROWLINSON, J. S. (1970), "Structure and Properties of Simple Liquids and Solutions: a Review", *Disc. Faraday Soc.,* 49, 30–42.

** SCOTT, R. L. (1971), "Introduction", in *Physical Chemistry: An Advanced Treatise,* eds. Eyring, H., Henderson, D. and Jost, W., Vol. 8A, *Liquid State,* ed. Henderson, D., Academic Press, New York, Chap. 1, pp. 1–83.

** WINTERTON, R. H. S. (1970), "Van der Waals Forces", *Contemp. Phys.,* 11, 559–74.

Specific references

DAWE, R. G., MAITLAND, G. C., RIGBY, M. and SMITH, E. B. (1970), "High Temperature Viscosities and Intermolecular Forces of Quasi-spherical molecules", *Trans. Faraday Soc.,* 66, 1955–65.

HAMANN, S. D. and LAMBERT, J. A. (1954), "The Behaviour of Fluids of Quasi-Spherical Molecules", *Austral. J. Chem.,* 7, 1–17, 18–27.

HOOVER, W. G., GRAY, S. G. and JOHNSON, K. W. (1971), "Thermodynamic Properties of the Fluid and Solid Phases for Inverse-Power Potentials", *J. Chem. Phys.,* 55, 1128–36.

KAC, M. (1959), "On the Partition Function of a One-Dimensional Gas", *Phys. Fluids,* 2, 8–12.

KAC, M., UHLENBECK, G. E. and HEMMER, P. C. (1963), "On the van der Waals Theory of the Vapor–Liquid Equilibrium. I and II", *J. Math. Phys.*, 4, 216–28, 229–47.

KIHARA, T. (1953), "Virial Coefficients and Models of Molecules in Gases", *Rev. Mod. Phys.*, 25, 831–43.

VAN DER WAALS, J. D. (1873), *On the Continuity of the Gas and Liquid Phases,* Dissertation, Leiden; English translation, Threlfall and Adair (1890), *Physical Memoirs,* 1, 333.

Chapter 12
Theories of the liquid state

The full description of a system of molecules involves the specification of the momenta and the positions of all the molecules as a function of time. This problem is sometimes called the "many body" problem, and may be tackled in several ways. In principle it is possible to predict the equilibrium or thermodynamic properties of a liquid using only the experimental information contained in the effective pair potential energy function, but the prediction of liquid non-equilibrium or transport properties requires a description of the collision mechanism: a kinetic equation analogous to the Boltzmann equation in dilute gases. For this reason equilibrium and non-equilibrium theories in general have developed separately, but for present purposes they may be considered together under the following headings.

(i) Statistical thermodynamic distribution function theories

It is possible in principle, using statistical mechanics (Rice and Gray, 1965), to calculate analytically distribution functions, thermodynamic properties and transport coefficients for an (assumed known) effective pair potential energy function, but in practice this is extremely difficult, particularly for the non-equilibrium properties.

(a) Integral equation methods

Probably the most widely used equilibrium theories of liquids are those which aim to calculate the pair distribution function and the equilibrium or thermodynamic properties from an (assumed known) intermolecular potential energy function $\phi(r)$ in terms of *integral equations*. The pair distribution function is proportional to an average of the Boltzmann factor, $\exp(-\Phi/kT)$, for the energy of interaction of molecules 1 and 2 with all other $(L-2)$ molecules. The potential energy Φ of the configuration is usually defined on the assumption of pairwise additivity of interactions between molecules (Chapter 11). The averaging is done by integrating over all possible positions for molecules 1 and 2, but the problem is the evaluation of the integrals: this is not possible for dense fluids without further approximations.

One approach to the problem is to differentiate the integral equation with respect to the separation r_{12} of the two representative molecules 1 and 2. The reason for adopting this procedure is that the *deviation* of $g^{(2)}(r)$ about its equilibrium value in a fluid structure would be expected to be much more weakly dependent than the *actual nature* of $g^{(2)}(r)$ on details of the interaction potential (which are not known very well). The resulting equation is still exact (subject only to the assumption of pairwise interaction) but it relates $g^{(2)}(r)$ to $\phi(r)$ through another unknown function, $g^{(3)}(r)$, which was introduced in Chapter 5. It is then possible to make the *superposition approximation* proposed by J. G. Kirkwood, on the basis that the chance of the occurrence of a group of three molecules in a fluid can be expected to be closely related to the product of the probabilities of the occurrence of the three separate pairs:

$$g^{(3)}(r) = g^{(2)}(\mathbf{r}_1, \mathbf{r}_2)g^{(2)}(\mathbf{r}_1, \mathbf{r}_3)g^{(2)}(\mathbf{r}_2, \mathbf{r}_3)$$

(It may be noted here that for fluids with simplified interaction potentials such as the hard sphere and Lennard–Jones type (Chapter 11) where its validity can be tested by computer simulation (see below) the superposition approximation leads to an over-estimation of the degree of order.) Using this approximation there is obtained the *Born–Green* integral equation, a non-linear integral equation connecting $g^{(2)}(r)$ and $\phi(r)$ from which may be obtained by computer calculation the equilibrium properties of fluids, at least for simple forms of $\phi(r)$.

An alternative but related approach is the *virial expansion* of correlation functions, analogous to the virial equation (2.4). Because the direct correlation function $c(r)$ decreases rapidly as r decreases (Fig. 5.8c) its range does not greatly exceed that of $\phi(r)$. For this reason the expansion of $c(r)$ has been widely used:

$$c(r) = A + B\rho + C\rho^2/2 + \cdots$$

where the coefficients A, B, C, ... are called *cluster integrals*. Various methods of terminating the series expansion give rise to the "simple chain", "netted chain" and "hyper-netted chain" (HNC) approximations.

Percus and Yevick (1958) introduced a new approach by considering the differences between liquids and *solids* as a function of density rather than by using a series expansion starting from the perfect *gas*. The resulting ("PY") equation is, however, rather similar to the virial expansion approximations, with a number of cluster integral terms intermediate between those of the netted chain and the HNC equations. The PY equation is numerically superior to the HNC equation for a hard sphere fluid: the inclusion of a greater number of terms in the approximation does not necessarily provide greater numerical accuracy.

Improvements to the PY and HNC equations have included the incorporation of triplet potential energies:

$$\Phi = \sum_{1 \leq i < j \leq L} \phi(i, j) \; + \; \sum_{1 \leq i < j < k \leq L} \phi(i, j, k)$$

These have been reviewed, for example, by Rushbrooke (1968) and McDonald and Singer (1970b). Several methods have been developed to solve the equations, involving iteration and Fourier transformation (de Boer *et al.*, 1964; Watts, 1968). Numerical tests of these integral methods have been described by Rowlinson (1968) and Watts (1971).

(b) Transport properties

Statistical mechanical theories of transport processes in fluids have also been developed. Kirkwood (1946) used Brownian motion concepts and the Langevin equation describing the motion of a large particle in dilute solution. This involved the approximation that the force exerted on a molecule in a dense fluid contains two parts, one due to the fluctuating environment and one proportional to the momentum of the molecule. A time interval τ was postulated, long enough for averages of the *fluctuating* intermolecular forces over successive τ increments to follow the statistics of independent events, but short enough that a large *momentum* transfer was unlikely to occur. This implied that the diffusion of a molecule involved a sequence of small steps with frequency much greater than $1/\tau$. It was possible to obtain a molecular friction coefficient ξ_i (compare with the friction coefficient ζ_i defined in equation (3.11)) in terms of an integral over the autocorrelation of the total intermolecular force $\mathbf{F}_i$ on molecule i during the time interval τ:

$$\xi_i = \frac{1}{3kT} \int_0^\tau < \mathbf{F}_i(0) \cdot \mathbf{F}_i(t) > \mathrm{d}t$$

(Autocorrelation functions were introduced at the end of Chapter 5.)

In order to account for the large momentum transfers that must occur during strongly repulsive encounters between neighbouring molecules, Rice and Allnatt (1961) modified Kirkwood's theory by dividing the intermolecular forces into components associated with (i) the short-range (rigid core) pair potential, and (ii) the long-range (soft, attractive at long distances and repulsive at shorter distances) pair potentials. The basic dynamic event was considered to consist of a repulsive binary collision followed by quasi-Brownian destruction of correlation in the rapidly fluctuating soft force field of the surrounding molecules. (A similar situation is illustrated from a different point of view in Fig. 7.6.) An expression for the doublet distribution function was derived, together with expressions for the transport coefficients such as viscosity and thermal conductivity. The agreement between theory and experiment for simple liquids suggests that the Kirkwood–Rice–Allnatt theory is a reasonable first-order description of transport in simple liquids on a molecular level.

(c) Fluctuation dissipation theory

Kinetic theories of dense fluids show that an irreversible approach to equilibrium is compatible with dynamic microscopic reversibility. The theory which describes irreversible (dissipative) systems in terms of reversible microscopic fluctuations is called the fluctuation–dissipation theory (Zwanzig, 1965; Egelstaff, 1966; Kubo, 1966). The macroscopic transport coefficients have been shown to be related to the autocorrelation function for fluctuation of an appropriate quantity for a system in thermal equilibrium (in the absence of the force). This means that the same information is accessible from a study of the response to an applied force as is obtained by observing the average molecular fluctuations by a technique such as radiation scattering. For example, the diffusion coefficient D is related to the velocity autocorrelation function by

$$D = \int_0^\infty \langle v_x(0) \cdot v_x(t) \rangle \, dt \tag{12.1}$$

where v_x is the component of the velocity in the direction being measured. Reference to Fig. 5.7 and equation (12.1) shows that the diffusion coefficient D may be evaluated if the velocity autocorrelation function is known, and also that the effect of negative regions in the velocity autocorrelation function is to reduce the value of D, as would be expected intuitively from the "cage" effect. The value of D for liquid argon is established within a time of the order of 10^{-12} s. Harris (1972) has pointed out that equation (12.1) may be replaced by a force correlation function expression related to the expression quoted for ξ, above:

$$D \propto \int_0^\infty t^2 \langle \mathbf{F}(0)\mathbf{F}(t) \rangle \, dt$$

(ii) Perturbation methods

It is assumed that the major features of the structure of a liquid are described by a system of hard spheres and that the details of the attractive and repulsive regions in the effective pair potential energy of real systems can be treated as small perturbations on the hard sphere potential.

For example (Zwanzig, 1954: Watts, 1971), the intermolecular potential energy can be written as the sum of the hard sphere potential energy $\phi_0(r)$ and a modifying energy function:

$$\phi(r) = \phi_0(r) + \varepsilon\psi(r)$$

$$\varepsilon\psi(r) = \begin{cases} 0 & r < \sigma \\ 4\varepsilon\left[\left(\dfrac{\sigma}{r}\right)^{12} - \left(\dfrac{\sigma}{r}\right)^{6}\right] & r \geqslant \sigma \end{cases}$$

where the symbols have the same meaning as in equation (11.12). This may be considered a first order perturbation theory; subsequent improvement by Barker and Henderson (1967) introduced two parameters, α and γ, in such a way that when $\alpha = \gamma = 0$ a hard sphere potential was obtained, and for $\alpha = \gamma = 1$ the original (first order perturbation) interaction potential was recovered. In this second order perturbation treatment α was used to control the "steepness" of the positive potential energy and γ was varied to alter the depth of the negative well. The Barker–Henderson theory was found to give internal energy and pressure results for Lennard–Jones (12, 6) potential fluids which agreed well with computer simulation and with physical experimental data from argon.

(iii) Structural or lattice theories

In addition to the more or less rigorous methods (i) and (ii), other approaches to the problem of liquid thermodynamic and transport properties have been made on the basis of various structural theories. A physical model is suggested which is a compromise between reality and mathematical tractability. This is then converted into a mathematical equation from which the properties of the liquid may be calculated. Comparison of these calculated values with experimental values permits a judgment to be made on the validity or otherwise of the initial model. It was reported in Chapter 5 that evidence points to the existence of "empty space" in liquids, and basically the various models differ in the way in which they describe this empty space. Liquids can be considered to arise either from the melting of a solid lattice or from the condensation of a gas. These two possibilities lead to solid-based and gas-based model structures for liquids. However, even in some gas-based models there comes a point where an appeal is made to some aspect of lattice geometry, hence the general description "lattice theories" for liquid state structural theories.

A disadvantage of the "lattice" approaches is the introduction of simplifications based on considering the liquid divided up into a number of "cells", (e.g. formed by shells of neighbouring molecules) arranged regularly on a lattice, each containing one molecule with a certain degree of freedom of movement. This approximation has had some successes, but has overemphasized the regularity in liquid systems and the similarity to the solid state. Considerable refinements have been made, including allowance for vacant and multiply-occupied cells and although the thermodynamic properties of liquids can be accounted for reasonably well, most of these theories cannot account for the fluidity of liquids. However, one structural theory to receive considerable attention during the last ten years is the *significant structures theory* (Eyring and Marchi, 1963; Eyring and Jhon, 1969; Jhon and Eyring, 1971).

It is assumed in the significant structures approach that the translational motions of molecules may be of two types: *solid-like*, for molecules moving in a potential energy well formed by their neighbours; and *gas-like*, for molecules moving in fluidized voids in the liquid. The "significant structures" are the porous, pseudo-crystalline framework, and the set of molecules occupying the voids in this framework. It is assumed that there are just enough molecules in a certain volume of vapour to fill the voids in the same volume of liquid (see Fig. 12.1). A void converts what would be otherwise three *vibrational* degrees of freedom of its neighbours into three *translational* degrees of freedom, i.e. it confers gas-like properties on the neighbouring molecules. Considerations such as these lead to an expression from which the various properties (thermodynamics, transport, surface, dielectric) of liquids may be evaluated. For simple liquids several parameters in the expression are obtained from solid state properties, and the results are surprisingly good in view of the purely intuitive basis of this method. Support for this model is provided by pair distribution function results (Chapter 5).

The fluidity of liquid arises because there is a variety of different structural states available, so that a particular state can relax into an alternative structure in response to a stress. It may be that the best approach to transport properties is through the *distribution* of structure rather than its average. On the basis of physical models of liquids (Chapter 10) Bernal (1959, 1960; Bernal and Mason, 1960) proposed that liquids could be described as homogeneous, coherent, and essentially irregular assemblies characterized by a distribution of coordination number, but this idea was not developed quantitatively because the statistical geometry needed to estimate the distribution of coordination number was not known. This method has been extended (Barton and Speedy, 1974) by relating the local microstructure to transport processes and assuming that structural rearrangement occurs by a mechanism involving uncoordinated molecules. With the same idea of structural distribution, the theory of Adam and Gibbs (1965) relates transport to the structural variety through the *configurational entropy* which can be evaluated from the thermodynamic properties of a material in crystalline and amorphous states.

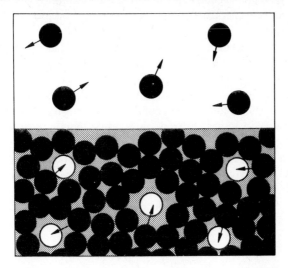

Fig. 12.1 Diagram depicting a liquid formed by the removal of molecules at random from a solid and fluidization of the vacancies. The behaviour of the fluidized vacancies mirrors the behaviour that a gas comprised of the missing molecules would show (after Eyring *et al.*, 1963).

(iv) Computer simulation*

Molecules in a real liquid can be considered to determine their arrangements by trying all possibilities. If one has adequate computing facilities it is possible to make a direct attack on the "many-body" problem by doing the same but with considerably fewer molecules. A mathematical model is set up in a hypothetical cubic box, with a limited number (of the order of hundreds) of "molecules" with an effective pair potential energy function of a suitable simplicity. The thermodynamic properties of this model can then be determined by this *numerical* method and the results compared with theoretical calculations of the types (i)–(iii) above on molecules with the same simplified properties. Consequently computer simulation can be considered as "experiments" carried out on liquids made up of artificially simple molecules interacting in an artificially simple

* Wood, 1968; McDonald and Singer, 1970a, 1973.

way. To minimize surface effects a "periodic boundary condition" is used, so that the whole of space is imagined to be filled with periodic reproductions of the basic "cell": as a molecule leaves through one wall of the cube it enters through the opposite wall. One limitation of computer methods used up to the present because of the great simplification which it introduces has been the pairwise additivity assumption. Nevertheless this development has stimulated the application of theoretical approaches to these hypothetical liquid systems because the results can now be tested.

In the *Monte Carlo* method (Metropolis *et al.*, 1953) the particles are displaced one at a time according to rules which ensure that in the successive configurations, individual configurations appear with probability proportional to the Boltzmann factor, $\exp(-\Phi/kT)$ where Φ is the potential energy of the configuration. The thermodynamic energy can be calculated by taking an average over these configurations, which because temperature, volume and composition are fixed is over a "canonical ensemble" in the terminology of statistical thermodynamics,

$$U = (L/N)\langle \Phi \rangle \tag{12.2}$$

where L is the Avogadro constant and N is the number of molecules considered. The Monte Carlo method provides information on equilibrium properties only, and not on transport properties because the moves of the molecules are not associated with real time.

In the *molecular dynamics* method (Alder and Wainwright, 1959) the molecules, although they may be given initially equal kinetic energies, have specified velocities in random directions, so successive configurations may be found by solving to the required accuracy the simultaneous Newtonian equations of motion after successive short periods of time. The volume, composition and *total energy* rather than temperature are fixed in this method: it is a "microcanonical ensemble" that is modelled, and the temperature of the system varies from instant to instant as the *kinetic* energy varies. The temperature and other thermodynamic properties are calculated using a *time* average (whereas thermodynamic properties follow from an ensemble average in the Monte Carlo method) and in consequence the molecular dynamics method permits the calculation of transport properties. More importantly, the molecular dynamics method can provide a series of "snapshots", showing how a particular molecular distribution develops in time. Therefore although this technique is sometimes criticized on the basis of its artificially simple molecular interactions, it is unique among experimental techniques and theoretical calculations in providing information on fundamental, individual molecular configurations.

As an example of a typical molecular dynamics experiment, it is interesting to consider the results of calculations on a two-dimensional system of Lennard–Jones discs (Fehder, 1969; Emeis and Fehder, 1970). The data were displayed graphically both as "snapshots" of the system configuration at a particular instant of time (Fig. 12.2) and as plots of the trajectories of particle centres over

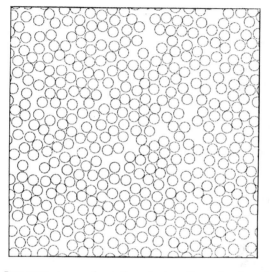

Fig. 12.2 Instantaneous configuration of a two-dimensional model fluid in a liquid state determined by molecular dynamics calculations. The particles are plotted with a diameter σ, the Lennard–Jones function distance parameter (equation 11.12) (after Emeis and Fehder, 1970).

various time intervals (Fig. 12.3). The snapshots show relatively large vacancies, and the trajectory plots that these vacancies can persist in the same region for relatively long periods ($> 10^{-12}$ s).

Computer calculations of the properties of random assemblies of hard spheres (Barker and Henderson, 1971b; Finney, 1971b; Adams and Matheson, 1972) are found to agree with the measurements on physical models discussed in Chapter 10. Computer simulation experiments can be used to predict the occurrence of phase transitions in systems of particles with simple effective pair potential energy functions. The hard sphere model in three dimensions and the hard disc model in two dimensions have one phase

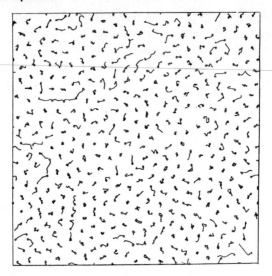

Fig. 12.3 Trajectories of the particles in a liquid-like state of a two-dimensional model fluid determined by molecular dynamics calculations. The small circles mark the initial positions of the particles (corresponding to the configuration in Fig. 12.2) and the irregular lines denote the paths of the centres during a 2×10^{-12} s interval (after Emeis and Fehder, 1970).

transition between a disordered or fluid phase at low densities and an ordered or solid phase at high densities; i.e. a hard sphere or hard disc fluid behaves as a substance above its critical temperature. The square well potential function model (Fig. 11.2) in two and three dimensions is the simplest model to give rise to all three states of matter: an attractive force is essential for the existence of a liquid state.

The "simplest" liquid, argon, has been simulated by computer (Rahman, 1964; Barker *et al.*, 1971), and Woodcock and Singer have extended computer simulation to molten salts, showing that this technique is practicable for ionic liquids of high charge density by Monte Carlo (Woodcock and Singer, 1971; Woodcock, 1971b) and molecular dynamics (Woodcock, 1971a, 1972) methods. The effective pair potential energy function used was

$$\phi(r) = ar^{-1} + b \exp [c(\sigma - r)] + dr^{-6} + er^{-8} \qquad (12.3)$$

This has terms representing coulomb potential, short-range repulsion, dipole–dipole, and dipole–quadrupole energies. Woodcock

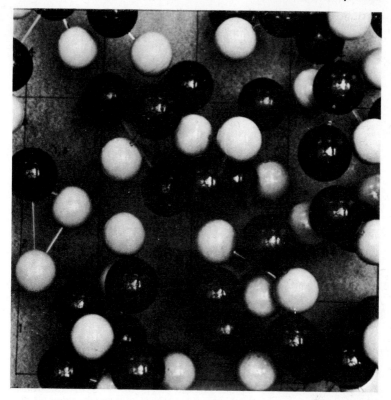

Fig. 12.4 Model of the arrangement of ions in molten potassium chloride from a molecular dynamics calculation (from Woodcock, 1971b).

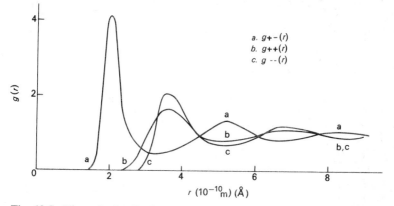

Fig. 12.5 The pair distribution functions in molten lithium chloride from molecular dynamics calculations for (a) cation–anion, (b) cation–cation, and (c) anion–anion pairs (after Woodcock, 1971a).

(1971b) has pointed out that computer simulation provides a means for generating realistic liquid configurations which can be used in model construction. Figure 12.4 shows a model of the geometry of ions in molten potassium chloride as indicated by a computer generated configuration. Additional currently experimentally in-accessible information is also provided; for example, the pair distribution functions for cation–cation, cation–anion and anion–anion pairs (Fig. 12.5).

General references

★★★ ALDER, B. J. and HOOVER, W. G. (1968), "Numerical Statistical Mechanics", in Temperley *et al.* (1968), Chap. 4, pp. 79–113.

★★★ BARKER, J. A. (1963), *Lattice Theories of the Liquid State,* Pergamon, Oxford.

★★★ BARKER, J. A. and HENDERSON, D. (1971a), "Some Aspects of the Theory of Liquids", *Accounts Chem. Res., 4,* 303–7.

★★★ BARKER, J. A. and HENDERSON, D. (1972), "Theories of Liquids", *Ann. Rev. Phys. Chem., 23,* 439–84.

★★★ BAXTER, R. J. (1971), "Distribution Functions", in Henderson (1971), Vol. 8A, Chap. 4, pp. 267–334.

★★★ BERNE, B. J. (1971), "Time-Dependent Properties of Condensed Media", in Henderson (1971), Vol. 8B, Chap. 9, pp. 539–716.

★★★ BERNE, B. J. and FORSTER, D. (1971), "Topics in Time-Dependent Statistical Mechanics", *Ann. Rev. Phys. Chem., 22,* 563–96.

★★★ COLE, G. H. A. (1971), "Statistical Theories of the Liquid State", in *Essays in Physics,* eds. Conn, G. K. T. and Fowler, G. N., Academic Press, London, Vol. 3, pp. 35–116.

★★ CRUICKSHANK, A. J. B. (1971), "Nonelectrolyte Liquids and Solutions", in *Problems in Thermodynamics and Statistical Physics,* ed. Landsberg, P. T., Pion, London, pp. 140–192.
Includes in the form of problems and solutions the configurational thermodynamic approach to liquids.

★ DREISBACH, D. (1966), *Liquids and Solutions,* Houghton Mifflin, Boston, U.S.A.

★★ EGELSTAFF, P. A. (1967), *An Introduction to the Liquid State,* Academic Press, London, Chap. 5.

*** ENDERBY, J. E. (1972), "The Correlation Function for Simple Liquids", *Adv. Struct. Res. Diffr. Meth.,* 4, 65–104.

 * EYRING, H. and MARCHI, R. P. (1963), "Significant Structure Theory of Liquids", *J. Chem. Educ.,* 40, 562–72.

 * EYRING, H., HILDEBRAND, J. and RICE, S. (1963), "The Liquid State", *Internat. Sci. and Technology,* No. 15, 56–66.
An interesting introduction to liquid state theory.

 * FLOWERS, B. H. and MENDOZA, E. (1970), *Properties of Matter,* Wiley, London, pp. 211–17, 297–304.

** FLUENDY, M. A. D. and SMITH, E. B. (1962), "The Application of Monte Carlo Methods to Physicochemical Problems", *Quart. Rev.,* 16, 241–66.

*** GRAY, P. (1968), "The Kinetic Theory of Transport Phenomena in Simple Liquids", in Temperley *et al.* (1968), Chap. 12, pp. 507–62.

** HENDERSON, D. (1971), ed., *Liquid State,* in *Physical Chemistry: An Advanced Treatise,* eds. Eyring, H., Henderson, D., and Jost, W., Vols. 8A and 8B, Academic Press, New York.
A collection of chapters at a moderately advanced level with a variety of points of view.

** HENDERSON, D. and BARKER, J. A. (1971), "Perturbation Theories", in Henderson (1971), Vol. 8A, Chap. 5, pp. 377–412.

*** HENDERSON, D. and DAVISON, S. G. (1967), "Equilibrium Theory of Liquids and Liquid Mixtures", in *Physical Chemistry: An Advanced Treatise,* eds. Eyring, H., Henderson, D. and Jost, W., Vol. 2, *Statistical Mechanics,* ed. Eyring, H., Academic Press, New York, Chap. 7, pp. 339–404.

*** HIRSCHFELDER, J. O., CURTISS, C. F. and BIRD, R. B. (1954), *Molecular Theory of Gases and Liquids,* Wiley, New York.
A "standard work" on gases and liquids from a molecular and statistical thermodynamic viewpoint.

** HUGHEL, T. J. (1965), ed., *Liquids: Structure, Properties, Solid Interactions,* Proc. Symposium, Warren, Michigan, 1963, Elsevier, Amsterdam.
Includes chapters on several aspects of liquid theory.

** JHON, M. S. and EYRING, H. (1971), "The Significant Structures Theory of Liquids", in Henderson (1971), Vol. 8A, Chap. 5, pp. 335–75.

★★★ KOHLER, F. (1972), "On the Theory of Simple Liquids", *Adv. Molec. Relaxation Processes,* 3, 297–314.

★★ LAL, M. (1971), "Computer Simulation of Some Physicochemical Problems", *R.I.C. Rev.,* 4, 97–128.

★★ MCDONALD, I. R. and SINGER, K. (1970a), "The Study of Simple Liquids by Computer Simulation", *Quart. Rev.,* 24, 238–62.

★★ MCDONALD, I. R. and SINGER, K. (1970b), "Theories of Dense Fluids", *Ann. Rep. Chem. Soc.,* 67A, 53–64.

★ MCDONALD, I. R. and SINGER, K. (1973), "Computer Experiments on Liquids", *Chem. in Britain,* 9, 54–60, 65.
A "current awareness" review.

★★★ MOUNTAIN, R. D. (1970), "Liquids: Dynamics of Liquid Structure", *Crit. Rev., Solid State Sci.,* 1, 5–46.

★★ NEECE, G. A. and WIDOM, B. (1969), "Theories of Liquids", *Ann. Rev. Phys. Chem.,* 20, 167–90.

★★ PRYDE, J. A. (1966), *The Liquid State,* Hutchinson, London.

★★★ REE, F. H. (1971), "Computer Calculations for Model Systems", in Henderson (1971), Vol. 8A, Chap. 3, pp. 157–266.

★★★ RICE, S. A. and GRAY, P. (1965), *The Statistical Mechanics of Simple Liquids,* Interscience, New York.

★★ ROWLINSON, J. S. (1964), "The Theory of Liquids at Equilibrium", *Contemp. Phys.,* 5, 359–76.

★★ ROWLINSON, J. S. (1965), "The Equation of State of Dense Systems", *Rpts Progr. Phys.,* 28, 169–99.

★★★ ROWLINSON, J. S. (1968), "A Comparison of Integral Equations for the Distribution Functions with the Properties of Model and Real Systems", in Temperley *et al.* (1968), Chap. 3, pp. 59–77.

★★ ROWLINSON, J. S. (1969), *Liquids and Liquid Mixtures,* Butterworths, London, 2nd ed., Chap. 8.

★★★ RUSHBROOKE, G. S. (1968), "Equilibrium Theories of the Liquid State", in Temperley *et al.* (1968), Chap. 2, pp. 25–58.

★ SMITH, B. L. (1971), *The Inert Gases: Model Systems for Science,* Wykeham, London, particularly Chap. 6, "Liquids".

★★★ TEMPERLEY, H. N. V., ROWLINSON, J. S. and RUSHBROOKE, G. S. (1968),

eds., *Physics of Simple Liquids,* North-Holland, Amsterdam.
A comprehensive collection of chapters at a rather advanced level.

** WATTS, R. O. (1971), "Recent Advances in the Theory of Liquids",
Rev. Pure Appl. Chem., 21, 167–88.

*** WOOD, W. W. (1968), "Monte Carlo Studies of Simple Liquid
Models", in Temperley *et al.* (1968), Chap. 5, pp. 115–230.

Specific references

ADAM, G. and GIBBS, J. H. (1965), "On the Temperature Dependence
of Co-operative Relaxation Properties in Glass-Forming Liquids",
J. Chem. Phys., 43, 139–46.

ADAMS, D. J. and MATHESON, A. J., "Cell Theory for the Liquid State",
J. Chem. Soc. Faraday II, 68, 1536–46.

ALDER, B. J. and WAINWRIGHT, T. W. (1959), "Phase Transition for
a Hard Sphere System", *J. Chem. Phys.,* 27, 1208–9.

BARKER, J. A., FISHER, R. A. and WATTS, R. O. (1971), "Liquid Argon.
Monte Carlo and Molecular Dynamics Calculations", *Molec.
Phys.,* 21, 657–73.

BARKER, J. A. and HENDERSON, D. (1967), "Perturbation Theory and
Equation of State for Fluids: I. The Square Well Potential; II.
A Successful Theory of Fluids", *J. Chem. Phys.,* 47, 2856–61,
4714–21.

BARKER, J. A. and HENDERSON, D. (1971b), "Monte Carlo Values for
the Radial Distribution Function of a System of Fluid Hard
Spheres", *Molec. Phys.,* 21, 187–91.

BARTON, A. F. M. and SPEEDY, R. J. (1974), "Simultaneous Conductance and Volume Measurements on Molten Salts at High
Pressure", *J.C.S. Faraday I,* 70, 506–27.

BERNAL, J. D. (1959), "A Geometrical Approach to the Structure of
Liquids", *Nature,* 183, 141–7.

BERNAL, J. D. (1960), "Geometry of the Structure of Monatomic
Liquids", *Nature,* 185, 68–70.

BERNAL, J. D. and MASON, J. (1960), "Coordination of Randomly
Packed Spheres", *Nature,* 188, 910–11.

DE BOER, J., VAN LEEUWAN, J. M. J. and GROENVELD, J. (1964), "Calculation of the Pair Correlation Function. II. Results for the 12–6 Lennard–Jones Interaction", *Physica,* 30, 2265–89.

EGELSTAFF, P. A. (1966), "Microscopic Transport Phenomena in Liquids", *Rpts. Progr. Phys.,* 29, 333–71.

EMEIS, C. A. and FEHDER, P. L. (1970), "The Microscopic Mechanism for Diffusion and the Rates of Diffusion-Controlled Reactions in Simple Liquid Solvents", *J. Amer. Chem. Soc.,* 92, 2246–52.

EYRING, H. and JHON, M. S. (1969), *Significant Liquid Structures,* Wiley, New York.

FEHDER, P. L. (1969), "Molecular Dynamics Studies of the Microscopic Properties of Dense Fluids", *J. Chem. Phys.,* 50, 2617–29.

FINNEY, J. L. (1970), "Random Packings and Structure of Simple Liquids", *Proc. Roy. Soc. (London),* A319, 495–507.

HARRIS, S. (1972), "Force Correlation Representation for the Self-Diffusion Coefficient", *Molec. Phys.,* 23, 861–5.

KIRKWOOD, J. G. (1946), "The Statistical Mechanical Theory of Transport Processes. I. General Theory", *J. Chem. Phys.,* 14, 180–201.

KUBO, P. A. (1966), "The Fluctuation–Dissipation Theorem", *Rpts. Progr. Phys.,* 29, 255–84.

METROPOLIS, N., ROSENBLUTH, A. W., ROSENBLUTH, M. N., TELLER, A. H. and TELLER, E. (1953), "Equation of State Calculations by Fast Computing Machines", *J. Chem. Phys.,* 21, 1087–92.

PERCUS, J. K. and YEVICK, G. J. (1958), "Analysis of Classical Statistical Mechanics by Means of Collective Co-ordinates", *Phys. Rev.,* 110, 1–13.

RAHMAN, A. (1964), "Correlations in the Motions of Atoms in Liquid Argon", *Phys. Rev.,* 110, A405–11.

RICE, S. A. and ALLNATT, A. R. (1961), "On the Kinetic Theory of Dense Fluids. VI. Singlet Distribution Function for Rigid Spheres with an Attractive Potential. VII. The Doublet Distribution Function for Rigid Spheres with an Attractive Potential", *J. Chem. Phys.,* 34, 2144–55, 2156–65.

WATTS, R. O. (1968), "Percus–Yevick Equation Applied to a Lennard–Jones Fluid", *J. Chem. Phys.*, 48, 50–5.

WOODCOCK, L. V. (1971a), "Isothermal Molecular Dynamics Calculations for Liquid Salts", *Chem. Phys. Letters,* 10, 257–61.

WOODCOCK, L. V. (1971b), "Model of the Instantaneous Structure of a Liquid", *Nature Phys. Sci.*, 232, 63–4.

WOODCOCK, L. V. (1972), "Some Quantitative Aspects of Ionic Melt Microstructure", *Proc. Roy. Soc.,* A328, 83–95.

WOODCOCK, L. V. and SINGER, K. (1971), "Thermodynamic and Structural Properties of Liquid Ionic Salts obtained by Monte Carlo Computation. I. Potassium Chloride", *Trans. Faraday Soc.,* 67, 12–30.

ZWANZIG, R. (1954), "High Temperature Equation of State by Perturbation Methods. I. Nonpolar Gases", *J. Chem. Phys.,* 22, 1420–6.

ZWANZIG, R. (1965), "Time-Correlation Functions and Transport Coefficients in Statistical Mechanics", *Ann. Rev. Phys. Chem.,* 16, 67–102.

Chapter 13
Simple liquids and real liquids

In the past, much of the theoretical interest in the liquid state has been concerned with "simple" liquids, i.e. those liquids for which it is assumed that the particle energies are pairwise additive and spherically symmetric, and has dealt with the idealized hard sphere or Lennard–Jones (12, 6) type of potential energy systems. Real liquids can never conform completely to the definition of "simple" liquids, but as previously indicated, liquid argon approximates to a model based on spherically symmetric pair potential energy functions.

It is the purpose of this section to compare simple or ideal liquids with the many real liquids. The important factors in such comparison are as follows:

(i) The degree of *physical asymmetry* of the molecules. For this purpose a molecule may be considered symmetrical if the inter-nuclear distances are much smaller than the mean molecular separation in the liquid, so that the forces between molecules act effectively through the centres of mass and there is thus no torque effect. The liquefied "noble" gases are, of course, the ideal examples of symmetry. (It should be noted, however, that in the light liquids – helium, and to a smaller extent, neon – there is departure from classical behaviour because of the "quantum effect", a large zero point energy.) In the extreme asymmetry case of some complex organic molecules, *liquid crystals* are formed: a packing together in parallel sheets or columns without loss of total mobility. Although the *glassy state* (Kauzmann, 1948; Angell, 1970; Faraday, 1972) is commonly observed in the case of large molecules, particularly polymers, it occurs in many other liquids particularly if short-time scale experiments are used, so "glass forming liquids" do not form a separate class of liquids.

(ii) The degree of polarity or *electrical asymmetry*. Polarity ranges from small dipole moment values in the hydrogen halides and methyl halides to extremely polar molecules such as water and ammonia. Some liquids such as carbon dioxide and acetylene have large quadrupole moments.

(iii) Any *dissociation* or *association*. The degree of dissociation into ions or self-ionization is extremely small in many covalent liquids, but is essentially complete in molten salts such as molten sodium chloride. The electrical conductivity is a useful criterion for distinguishing "covalent" from "ionic" liquids: usually greater than 10^{-1} S cm^{-1} for ionic liquids and less than 10^{-4} S cm^{-1} for covalent liquids. The conductivity of liquid metals is even higher, usually greater than 10^4 S cm^{-1}. Association is typified by hydrogen bonded liquids such as water and sulphuric acid, and results in relatively high boiling point, low fluidity, and long liquid temperature range.

Because of the importance of water, it is predictable that liquids are commonly described as "aqueous" or "non-aqueous". Not surprisingly, most discussions on "non-aqueous" liquids have concentrated on *solvent* behaviour (Waddington, 1965, 1969; Hildebrand *et al.*, 1970), and although the fundamental property of molecular interaction controls solution properties in the same way as it determines the state or "structure" of pure liquids, this extensive subject will not be discussed here. The ability of a liquid to act as a solvent depends on a variety of properties. Non-electrolyte solubility is closely linked to the internal pressure (Chapter 2) and the ionizing ability of a liquid (the tendency to form electrolyte solutions) depends primarily on its relative permittivity (Chapter 8).

Classification of liquids

Any classification of liquids introduces over-simplifications, but the following, which combines features of the classifications of Rowlinson (1969) and Brönsted (Davis, 1970), is a useful guide:

Monatomic liquids

(i) Liquefied noble gases such as argon.

Molecular liquids

(ii) Homonuclear or approximately homonuclear liquids such as nitrogen and carbon monoxide.

(iii) Globular molecular liquids such as methane and silicon hexafluoride.

(iv) Non-polar aprotic or "inert" organic liquids, notably the hydrocarbons.

121

   (v) Aprotic inorganic liquids
       (a) oxides
       (b) aprotic halogen compounds: oxyhalides, polar halides, halogens, interhalogens.
  (vi) Organic and inorganic protic liquids.
  (vii) Polar aprotic organic liquids.
 (viii) Large molecule liquids, including liquid polymers and liquid crystals.

Dissociated liquids

  (ix) Ionic liquids (molten salts)
       (a) inorganic
       (b) organic
   (x) Metallic liquids
  (xi) Semiconducting liquids

General references

** CHEN, S.-H. (1971), "Structure of Liquids", in *Physical Chemistry: An Advanced Treatise,* eds. Eyring, H., Henderson, D. and Jost, W., Academic Press, New York, Vol. 8A *Liquid State.* Chap. 2, pp. 85–156.

** HILDEBRAND, J. H., PRAUSNITZ, J. M. and SCOTT, R. L. (1970), *Regular and Related Solutions,* Van Nostrand Reinhold, New York.

** ROWLINSON, J. S. (1969), *Liquids and Liquid Mixtures,* Butterworths, London, 2nd ed.

 * WADDINGTON, T. C. (1965), ed., *Non-Aqueous Solvent Systems,* Academic Press, London.

 * WADDINGTON, T. C. (1969), *Non-Aqueous Solvents,* Nelson, London.

Specific references

ANGELL, C. A. (1970), "The Data Gap in Solution Chemistry: The Ideal Glass Transition Puzzle", *J. Chem. Educ.,* 47, 583–7.

DAVIS, M. M. (1970), "Brønsted Acid–Base Behaviour in 'Inert' Organic Solvents", in *The Chemistry of Nonaqueous Solvents,* ed. Lagowski, J. J., Academic Press, New York, Vol. 3, Chap. 1, pp. 1–135, esp. pp. 13–16.

FARADAY Symposium, Chemical Society (London) (1972), 6, *Molecular Motions in Amorphous Solids and Liquids*.

KAUZMANN, W. (1948), "Nature of the Glassy State and Behaviour of Liquids at Low Temperatures", *Chem. Rev.*, 43, 219–56.

Chapter 14
The properties of some important liquids

In this chapter the structures and properties of a selection of liquids of laboratory and industrial importance are discussed briefly. The order follows the classification given in Chapter 13, and numerical values of some physical constants are collected in Table 14.1. The boiling point and liquid range provide an indication of the extent of attractive forces in the liquid, either specific interactions of the hydrogen bond type or coulombic effects in ionic liquids. Although as pointed out in Chapter 4, the critical and triple point temperatures have a more fundamental importance, the boiling point and freezing point are of more practical significance in most applications. The viscosities are fairly uniform, but when this property has a high value, specific associative interactions are indicated, as in the glycols, for example. Relative permittivity provides a guide to ionizing ability as a solvent, but as seen in the aprotic amides, a high dipole moment or dipole length does not always infer a high relative permittivity. The relative permittivities of the protic amides should be noted: they are considerably higher than those of water and sulphuric acid. It is apparent from Table 14.1 that there is a very wide range of properties among liquids, so if a particular combination of melting point, relative permittivity, and self-ionization is required, there is an excellent selection of pure liquids available without even considering liquid mixtures.

Non polar aprotic organic liquids (iv)*

These liquids, also described as "inert", when used as solvents are called "indifferent", "non-dissociating", "non-ionizing" or "non-co-ordinating". The term "aprotic" is used here in its original sense of "protolytically indifferent". These liquids include the aliphatic and aromatic hydrocarbons, the halogen derivatives of higher hydrocarbons, and carbon disulphide, and they are characterized

---

* Davis, 1970.

Table 14.1 Liquid physical properties

(a) liquid range (°C) at atmospheric pressure
(b) freezing point (°C)
(c) boiling point (°C)
(d) relative permittivity (dielectric constant); temperature (°C)
(e) viscosity (cP); temperature (°C) (1 cP $= 10^{-3}$ Pa s
(f) dipole moment, Debye units (1 D $= 3.33564 \times 10^{-30}$ C m)
(g) dipole length, p/e (pm), $e = 1.60219 \times 10^{-19}$ C

References as indicated in text

	a	b	c	d	e	f	g
(i) argon, Ar	3	−189	−186				
(ii) nitrogen, N_2	16	−210	−196				
carbon monoxide, CO	16	−205	−191				
(iii) methane, CH_4	21	−183	−162				
carbon tetrafluoride, CF_4	135	−150	−15				
(iv) carbon tetrachloride, CCl_4	100	−23	77	2.2/25	0.9/25	0	0
benzene, C_6H_6	74	6	80	2.3/25	0.6/25	0	0
cyclohexane, C_6H_{12}	74	7	81	2.0/25	0.9/25		
toluene, $C_6H_5CH_3$	206	−95	111	2.4/25	0.6/25	0.3	6
carbon disulphide, CS_2	158	−112	46	2.6/20	0.4/25	0.06	1
(v) Aprotic inorganic liquids							
(a) liquid oxides							
CO_2		−57	[−89]				
SO_2	65	−75	−10	15/0	15/0 4/−10	1.6	33
NO_2	32	−11	21			0.3	6
N_2O_4	33	−12	21	2/18			
(b) aprotic halogen compounds							
$AsCl_3$	143	−13	130	13/17	1/20	1.6	33
$AsBr_3$	185	35	220	9/35	5/35		
$SbCl_3$	148	73	221	33/75	3/95		
$SbBr_3$	183	97	280	21/100	7/100		
$BiCl_3$	215	232	447		32/260		
nitrosyl chloride ClNO	59	−65	−6	20/−10	0.6/−30	1.8	37
phosphoryl chloride $POCl_3$	107	1	108	14/22	1/25		
F_2	32	−220	−188	1.5/−190			
Cl_2	67	−101	−34	2.0/−35	0.5/−34		
Br_2	66	−7	59	3.1/20	1/16		
I_2	69	114	183	11/118	2/116		

ICI	73	α27 β14	100		4/28		
ClF$_3$	95	−83	12	4.6/12	0.4/12		
BrF$_3$	117	9	126		2/25		
BrF$_5$	102	−61	41	8/25	0.6/24		
IF$_5$	92	9	101	36/35	2/25		
(vi) Protic liquids							
water, H$_2$O	100	0	100	78/25	0.9/25	1.9	40
methanol, CH$_3$OH	163	− 98	65	33/25	0.5/25	1.7	35
ethanol, CH$_3$CH$_2$OH	193	−115	78	25/25	1.1/25	1.7	35
ethylene glycol, CH$_2$OH CH$_2$OH	210	−13	197	41	16.9/25	2.3	48
glycerol, 1,2,3 propanetriol, CH$_2$OHCHOHCH$_2$OH	272	18	290		1500/20		
formic acid, HCOOH	93	8	101	56/25	1.4/30	1.4	29
acetic acid, CH$_3$COOH	101	17	118	6/18	1.0/30	0.8	17
propionic acid, CH$_3$CH$_2$COOH	162	− 21	141	3/25	1.0/30	0.6	12
n-butyric acid, CH$_3$CH$_2$CH$_2$COOH	168	− 5	163	3/25	1.4/30	0.6	12
trifluoroacetic acid, CF$_3$COOH	86	−15	71	8/20	0.6/20	2.3	48
hydrogen chloride, HCl	29	− 114	− 85	14/−114	0.5/−95	1.1	23
hydrogen bromide, HBr	20	− 87	− 67	7/−85	0.8/−67		
hydrogen iodide, HI	16	− 51	− 35	3/−50	1.4/−35		
hydrogen cyanide, HCN	39	−13	26	115	0.2/25		
hydrogen sulphide,						17/−78	
H$_2$S	25	− 85	− 60	9/−78	0.5/−80	0.8/−78	
ammonia, NH$_3$	45	− 78	− 33	23/−33	0.3/−34	1.5	31
hydrogen fluoride,				175/−73			
HF	103	− 83	20	84/0	0.3/0	1.8	37
sulphuric acid, H$_2$SO$_4$	280	10	290	100/25	25/25		
nitric acid, HNO$_3$	125	− 42	83	50/14	7/25	2.2	46
formamide, HCONH$_2$	208	3	211	110/25	3/25	3.7	77
N-methyl formamide (NMF), HCONHCH$_3$	185	−4	181	182/25	2/25	3.8	79

acetamide, CH$_3$CONH$_2$	140	81	221	59/83	2/81	3.4	71
N-methyl acetamide (NMA), CH$_3$CONCH$_3$	175	31	206	166/40	3/40	3.7	77
pyridine, C$_5$H$_5$N	158	− 42	116	12/25	0.8/30	2.1	44

(vii) Polar aprotic organic liquids

bromomethane, methyl bromide, CH$_3$Br	97	− 94	3	10/0			
bromoethane, ethyl bromide, CH$_3$CH$_2$Br	157	− 119	38	9/20	0.4/25	1.9	40
l-bromopropane, propyl bromide, CH$_3$CH$_2$CH$_2$Br	181	− 110	71	8/25	0.5/25	1.9	40
acetone, (CH$_3$)$_2$CO	151	− 95	56	21/25	0.3/25	2.9	60
acetyl chloride, CH$_3$COCl	165	− 113	52	16/20		2.4	50
acetyl bromide, CH$_3$COBr	174	− 97	77	17/20		2.4	50
benzoyl fluoride, C$_6$H$_5$COF	153	3	156	23/20			
benzoyl chloride, C$_6$H$_5$COCl	198	− 1	197	23/25		3.3	69
benzoyl bromide, C$_6$H$_5$COBr	207	8	215	21/25	1.8/25	3.4	71
dimethyl formamide (DMF), HCON(CH$_3$)$_2$	214	− 61	153	37/25	0.8/25	3.9	81
N,N-dimethyl acetamide (DMA or DMAC), CH$_3$CON(CH$_3$)$_2$	186	− 20	166	38/25	0.9/25	3.8	79
acetonitrile, CH$_3$CN	108	− 44	82	36/25	0.3/30	3.4	71
nitromethane, CH$_3$NO$_2$	130	− 29	101	36/30	0.6/30	3.6	75
nitrobenzene, C$_6$H$_5$NO$_2$	205	6	211	35/30	1.6/30	4.0	83
(di)methyl sulphoxide (DMSO), (CH$_3$)$_2$SO	171	18	189	47/25	2/25	4.0	83

tetramethylene sulphone, sulpholane, tetrahydrothiophen dioxide,	258	29	287	43/30	10/30	4.8	100
tetrahydrofuran	175	− 109	66	8/25	0.5/25	1.8	37
1,4-dioxan	89	12	101	2.2/25	1/25	0.5	10
ethylene carbonate	212	36	248	90/40	2/40	0.4	8

(viii) Ionic liquids (molten salts)

(a) inorganic

NaCl	657	808	1465	~ 2	2/800
KCl	635	772	1407	~ 2	1/800
AgBr	1103	430	1533	6	2/700
$LiNO_3$		254	*	2.5	3/350
$NaNO_3$		310	*	2	2/350

(b) organic

$(n\text{-hexyl})_4 N^+ BF_4^-$	91	*	6/200
$(n\text{-butyl})_4 N^+ BF_4^-$	162	*	5/200
pyridinium chloride	146	*	5/155

* decomposes

by relative permittivities as low as 2 and seldom greater than 10 (see Table 14.1).

Aprotic inorganic liquids (v)

These compounds form one of two classes of "covalent" liquids (as distinct from the almost completely ionized molten salts) which exhibit self-ionization, the other class being the protic liquids. The term "aprotic" here means "not containing hydrogen".

(a) Liquid oxides

These compounds are characterized by low relative permittivities and rather short liquid ranges at atmospheric pressure.

Dinitrogen tetroxide (Addison, 1967; Waddington, 1969) undergoes two types of self-dissociation. The first is well established, and occurs to the extent of 0.01 % at the freezing point ($-12°C$) and 0.1 % at the boiling point (21°C)

$$N_2O_4 \rightleftarrows NO_2 + NO_2$$

Self-ionization has been proposed:

$$N_2O_4 \rightleftarrows NO^+ + NO_3^-$$

If this does occur, the ions would be present as ion pairs because of the low relative permittivity (2) and the conductivity is very small, $< 3 \times 10^{-13}$ S cm^{-1}. Also possible is the self-ionization

$$N_2O_4 \rightleftarrows NO_2^+ + NO_2^-$$

but this suggestion has not been substantiated.

Sulphur dioxide (Waddington, 1965b, 1969; Burow, 1970) has a relative permittivity (~ 15) which although low compared with water is higher than that of N_2O_4, making it a better solvent for ionic compounds. It has the additional advantage that it can be handled at 0°C under its own vapour in sealed glass vessels.

(b) Aprotic halogen compounds

Oxyhalides and polar halides (Payne, 1965; Waddington, 1969) These are largely the Group V halides and oxyhalides, and those liquids which have been most useful as solvents, mainly because of stability and availability, are listed in Table 14.1. Self-ionization may be described:

$$2AsCl_3 \rightleftarrows AsCl_2^+ + AsCl_4^-$$

The electrical conductivity of the pure liquid is a measure of the degree of self-ionization (although this also reflects the purity) and the trend in conductivities is

$$PCl_3 < AsCl_3 < SbCl_3 < BiCl_3$$

Some of the higher molecular weight and higher melting halides, such as bismuth and mercuric halides, approach molten salts in properties and the extent of self-ionization may be controlled by varying the temperature and pressure. In nitrosyl chloride there is evidence for the ionization

$$NOCl \rightarrow NO^+ + Cl^-$$

and the NO^+ ion has a high mobility, analogous to the H^+ ion in water. Phosphoryl chloride has a long liquid range and a higher viscosity than NOCl, suggesting greater association.

Halogens and interhalogens (Sharpe, 1965; Waddington, 1969). The extreme chemical reactivity of these liquids has restricted their use as solvents and limited the study of their properties. Liquid iodine has received more attention than the other halogens, and there is evidence that the intermolecular attraction which occurs in the solid persists in the liquid state. The self-dissociation is probably best represented by

$$3I_2 \rightleftarrows I_3^+ + I_3^-$$

with a very low value for the ionic product $[I_3^+][I_3^-]$, although there may be more complex species involved.

Bromine trifluoride, one of the better known interhalogen liquids is so reactive that it explodes with water and most organic matter and it reacts with asbestos with incandescence. Self-ionization of the interhalogens may be represented by equations such as

$$3ICl \rightleftarrows I_2Cl^+ + ICl_2^-$$
or
$$2ICl \rightleftarrows I^+ + ICl_2^-$$

although the detailed natures of the ionic species have not yet been established.

Protic liquids (vi)*

These "protic" or "protonated" liquids range from the type classi-fied by Brönsted (Davis, 1970) as "amphiprotic" (i.e. both protogenic and protophilic) with high relative permittivity such as water, to liquids such as acetic acid with low relative permittivities and high relative acidities or to liquids such as pyridine with low relative permittivities and high relative basicities. As a general "rule of thumb", all organic compounds containing hydrogen not bonded to carbon fall within this category.

For present classification purposes, *all* these liquids may be con-

* Waddington, 1969.

sidered to be amphiprotic, i.e. one molecule acts as an acid and the other as a base. This is formally described by the equations

$$2HA \rightleftarrows H_2A^+ + A^-$$
or
$$3HA \rightleftarrows H_2A^+ + HA_2^-$$

The degree of autoprotolysis or self-ionization is usually more extensive if the bond to hydrogen has considerable polarity, so the order of decreasing autoprotolysis is

$$H_3PO_4 > H_2SO_4 > HF > CH_3COOH > H_2O > C_2H_5OH > NH_3$$

Quantitatively, the degree of self-ionization is best assessed by comparing the autoprotolysis constants, but often the electrical conductivities of the pure liquids are used. (As previously pointed out, this latter measurement is subject to impurity error for the weakly ionized liquids.)

Water*

The structure of water has been discussed at great length on many occasions, and no attempt will be made to review this information. Recently the molecular dynamics computer simulation method (Rahman and Stillinger, 1971) and a cell model approach (Weres and Rice, 1972) have been used with an effective pair potential energy function (A. Ben-Naim and F. H. Stillinger in Horne, 1972) which incorporates the tendency of water to form tetrahedral arrays of linear hydrogen bonds, which agrees with X-ray diffraction results, and also with the structure expected from the known electronic structure of the water monomer (Symons, 1972). It is interesting to note that at high temperatures and high pressures (> 4000 bar, $> 400°C$) supercritical water appears to be composed of non-hydrogen-bonded monomers (Maier and Franck, 1966; Walrafen, 1968).

Alcohols†

As in the case of water, the presence of the polar hydroxyl group gives rise to a hydrogen bonded structure, with the effect of giving the liquids high boiling points and long liquid ranges. It can be seen from Table 14.1 that glycerol has melting point and viscosity con-

---

* Eisenberg and Kauzmann, 1969; Horne, 1970; Horne, 1972; Franks 1972.
† Drago and Purcell, 1965.

siderably higher than those of ethanol and methanol (but a comparable liquid range).

Carboxylic acids *

Acetic acid is second only to ammonia with respect to the extent of investigation among non-aqueous solvents. In fact it is not a particularly good solvent for electrolytes (relative permittivity 6 at 18°C) and is characterized by dimerization and possibly higher polymerization. The dimerization and "self-ionization" may be formulated

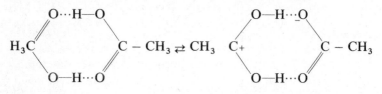

The higher relative permittivity of *formic acid* makes it a better electrolytic solvent, but it has the disadvantage of a tendency to decomposition unless kept under refrigeration. Nevertheless it deserves more attention than it has received. The higher homologues, propionic acid and *n*-butyric acid have longer liquid ranges and lower relative permittivities than acetic acid (Table 14.1). *Trifluoroacetic acid* is a very strong carboxylic acid and a very useful solvent. All these acids show evidence of association.

Higher hydrogen halides †

These three liquids (hydrogen chloride, hydrogen bromide and hydrogen iodide) resemble each other in physical properties and chemical behaviour, but differ appreciably from liquid hydrogen fluoride. The small electrical conductivity observed ($< 10^{-9}$ S cm^{-1}) is attributed to autoprotolysis

$$3HX \rightleftarrows H_2X^+ + HX_2^-$$

and the evidence from infra-red and nuclear magnetic resonance spectroscopic data favours the existence of a degree of hydrogen bonding in the liquid phase of these hydrogen halides, although this is very much weaker than in hydrogen fluoride. After pioneering studies of chemistry in these liquids in the first two decades of this

* Popov, 1970ab.
† Peach and Waddington, 1965; Klanberg, 1967.

century, further research was largely neglected until the late 1950s. The consequence of the low relative permittivities (3–9, see Table 14.1) is that only salts with low lattice energies (e.g. the quaternary ammonium halides) are soluble.

Hydrogen sulphide*

Liquid hydrogen sulphide has not received as much attention as some other non-aqueous liquids because of its physical and chemical properties: high toxicity, corrosive properties, difficulty of purification, low boiling point and narrow liquid range (although all these difficulties can be overcome if it is considered worthwhile). The H–S–H angle in the liquid state is close to 90°, and this is consistent with the negligible tendency to hydrogen bond formation: the free electron pairs are not localized in the more efficient sp^3 bonding orbitals as they are in water (H–O–H angle, 105°). This, of course, explains the relatively low boiling point of H_2S, determined by the van der Waals and general dipole–dipole forces.

Ammonia†

Liquid ammonia has been studied probably more than any other non-aqueous liquid. Like water, ammonia shows evidence of considerable hydrogen bonding. Because of its moderate relative permittivity, high dipole moment, hydrogen bonding ability and relatively high basicity, liquid ammonia is a very valuable solvent, particularly for non-electrolytes that are relatively insoluble in water.

Hydrogen fluoride ‡

Like water, hydrogen fluoride has a relatively high boiling point, a long liquid range, high relative permittivity and low entropy change on fusion, which suggest that it is an associated liquid. However, the relatively low surface tension ($10 \, \text{dyn cm}^{-1}$, $10^8 \, \text{N m}^{-1}$, at 0°C) and viscosity are not consistent with a three-dimensional structure as found in water, and a one- or two-dimensional associated system of long chains or perhaps cyclic hexamers (which occur in the gaseous state) are indicated, right up to the boiling point.

* Fehér, 1970.
† Jolly and Hallada, 1965; Lagowski and Moczygemba, 1967; Waddington, 1969.
‡ Hyman and Katz, 1965; Kilpatrick and Jones, 1967; Dove and Clifford, 1971.

Sulphuric acid*

This is probably the most widely studied of the strongly acidic solvents as it has a high relative permittivity and can be handled in glass apparatus at room temperature. Its high viscosity, high boiling point and high surface tension indicate a strongly associated, hydrogen bonded structure. In addition to autoprotolysis,

$$2H_2SO_4 \rightleftarrows H_3SO_4^+ + HSO_4^-,$$

sulphuric acid undergoes other types of self-dissociation and dehydration:

$$2H_2SO_4 \rightleftarrows H_2O + H_2S_2O_7$$
$$H_2O + H_2SO_4 \rightleftarrows H_3O^+ + HSO_4^-$$
$$H_2S_2O_7 + H_2SO_4 \rightleftarrows H_3SO_4^+ + HS_2O_7^-$$

Therefore, although sulphuric acid is an extremely valuable ionizing solvent, as a liquid it is particularly complex.

Nitric acid†

Nitric acid undergoes self-dissociation to a greater extent than any other pure liquid, and autoprotolysis is accompanied by dehydration:

$$2HNO_3 \rightleftarrows H_2NO_3^+ + NO_3^-$$
$$H_2NO_3^+ \rightleftarrows H_2O + NO_2^+$$

Because of this extensive self-dissociation (over 0.2 molal in NO_2^+ and NO_3^-) the electrical conductivity of nitric acid is one of the highest occurring in pure liquids other than liquid metals and molten salts. For the same reason nitric acid is only moderately strong as an acid. It is a strong oxidizing and nitrating agent, and much of the recent research on the properties of pure nitric acid was carried out in conjunction with nitration kinetic studies.

Protic amides ‡

The amide molecules have large dipole moments, but an amido hydrogen is required for a high relative permittivity (compounds

* Gillespie and Robinson, 1965; Lee, 1967a; Waddington, 1969.
† Lee, 1967b.
‡ Vaughn, 1967.

with high relative permittivity always have large dipole moments, but the reverse is not necessarily true). Thus the high relative permittivity amides with an amido hydrogen (notably formamide, *N*-methyl formamide, acetamide and *N*-methyl acetamide) fall in this class. The low relative permittivity dimethylformamide and dimethylaceta- mide belong to the polar aprotic organic liquid group. *Formamide* is thermally and photochemically unstable, and despite its similarity to water in properties affecting its solvent behaviour it has received relatively little attention. With one amide hydrogen replaced by a methyl group, *N-methyl formamide* (NMF) has superior stability and solvent properties. *Acetamide* has the lowest relative permittivity of the protic amides listed in Table 14.1, but is still a good ionizing solvent. It has also been used as a solvent for carbohydrates. *N- methyl acetamide* (NMA) has a lower melting point and a higher relative permittivity than acetamide, and consequently has received more attention as a solvent. The relatively high viscosity of the protic amides indicates a chain-like structure in these liquids.

Polar aprotic organic liquids (vii)[*]

This class comprises liquids which, although they contain hydrogen atoms cannot donate suitably labile hydrogen atoms to form strong hydrogen bonds, have high enough relative permittivities to allow study of ionic solutions (Parker, 1962; Ritchie, 1969). The extent of solvation of ions by polar aprotic liquids follows a very dif- ferent pattern from that which occurs in protic solvents. The suggested order of decreasing ability to solvate cations is DMSO, DMA > DMF > sulpholane $\gg$ CH_3CN, CH_3NO_2 > $C_6H_5NO_2$.

Dimethyl sulphoxide (DMSO)[†]

DMSO is more basic than acetone with respect to proton transfer,

$$H_3O^+ + (CH_3)_2SO \rightleftarrows H_2O + (CH_3)_2SOH^+$$

and to hydrogen bond formation, but as the autoprotolysis constant and acid dissociation constant values for DMSO are very small, this liquid is included here rather than in the "protonic liquid" class. DMSO possesses a hydrogen bonded structure which breaks down between 40 and 60°C, resembling water which also shows rapid structural changes as the temperature varies.

---

[*] Drago and Purcell, 1965; Davis, 1970.
[†] Reynolds, 1970.

Aprotic amides★

The contrast in relative permittivity, viscosity, melting point and structure between the protic amides discussed in (vi) and the aprotic amides is apparent from Table 14.1. *Dimethylacetamide* (DMA) has proved a good industrial solvent for polymers and co-polymers, and *dimethyl formamide* (DMF) is easily purified and a good solvent for both inorganic and organic compounds.

Acyl halides have been reviewed by Paul and Sandhu (1970).

Liquid crystals (viii) †

Some solids whose molecules are relatively long melt sharply to form a turbid, anisotropic liquid which on further heating transforms to a clear isotropic liquid. In some substances, particularly soaps, several different phases occur between the crystal and the isotropic liquid state. Figure 14.1 illustrates the possible degrees of order in long-molecule systems. The intermediate "liquid crystal" states

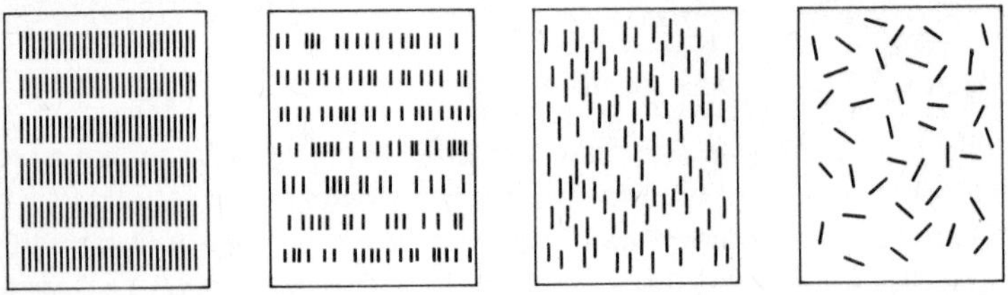

(a)   (b)   (c)   (d)

Fig. 14.1 Possible degrees of order in condensed states of long chain molecules: (a), crystalline - orientation and periodicity; (b), smectic – orientation and arrangement in equi-spaced planes, but no periodicity within planes; (c), nematic – orientation without periodicity; (d), isotropic fluid – neither orientation nor periodicity (after Moore, 1972).

(b) and (c) are characterized by disruption of the binding between the ends of the molecules while lateral attractions between the chains remain relatively strong. In the "smectic" state (named from the Greek word for "soap") the molecules lie in well-defined planes, and the "nematic" state (Greek for "thread") retains orientation but loses planar structure.

Liquid crystals are important both because they occur widely in living cells, and because of their potential use in electrically operated display devices.

★ Vaughn, 1967.
† Johnson and Porter, 1970; Moore, 1972; Kelker, to be published.

Ionic liquids (ix)*

Most molten salts resemble water in appearance, and near their melting points have viscosities, thermal conductivities and surface tensions which are similar to those of water. (Here water is used as a standard for comparison only because of its familiarity, not for any fundamental reason.) The liquid densities of the higher melting, inorganic salts are about 50% higher than that of water, but molten organic salts such as the quaternary ammonium halides have densities similar to that of water. It is in the property of electrical conductivity that molten salts differ markedly from other pure liquids: the conductivity of liquid sodium chloride at its melting point is $3.6\,\mathrm{S\,cm^{-1}}$, which is much greater than that of water $(4 \times 10^{-8}\,\mathrm{S\,cm^{-1}}$ at $25°C)$ and higher than that of a $1\,\mathrm{mol\,dm^{-3}}$ aqueous sodium chloride solution. Such salts are considered to be fully ionized. The conductivity is, however, much *lower* than that of a liquid metal (e.g. mercury, $1 \times 10^4\,\mathrm{S\,cm^{-1}}$ at $20°C)$, and is due to conductance by ions, not by electrons as in a metal.

Most salts of inorganic monatomic ions (notably the halides of alkali and alkaline earth metals) have been studied in the liquid state. There is a great variety of polyatomic inorganic ions, particularly oxyanion systems such as the nitrates. In the case of polyatomic ions, consideration must be given to chemical stability, because in addition to decomposition being faster at higher temperatures, it is to be expected that the strong, incompletely compensated local electric fields occurring in ionic melts would be associated with faster decomposition rates than occur in ionic crystals where crystal symmetry to some extent has a compensating effect (Ubbelohde, 1972). In general the lower melting inorganic salts exhibit some covalent bonding (incomplete dissociation) in the liquid state, and may alternatively be described as partially self-ionized covalent liquids (see e.g. class (v) (a)).

Salts of organic compounds provide the opportunity for making small changes of structure within a type of molecule, but for these systems the problem of chemical decomposition is a severe limitation. Generally, organic salts have greater temperature ranges of stability the lower the melting point, i.e. the stability is essentially independent of the melting point. Typical molten organic salts which have reasonable stability are the tetraalkylammonium halides, $R_4N^+X^-$; pyridinium halides, and alkali metal salts of carboxylic acids.

* Sundheim, 1964; Blander, 1964; Bloom, 1967; Janz, 1967; Bockris and Reddy, 1970; Rhodes, 1972; Clarke and Hills, 1973.

(Lind *et al.*, 1966; Reinsborough, 1968, Lind, 1973.)

It was pointed out in Chapter 12 that computer simulation methods have been applied to molten salts, and these liquids have also received considerable attention in the application of structural theories.

Metallic liquids (x)*

Up to this point it has been assumed that particle–particle interactions are the same in gaseous, liquid and solid states; that is, the total potential energy of a liquid has been expressed as a sum of effective pair potential energies. In a metal, the potential energy is made up

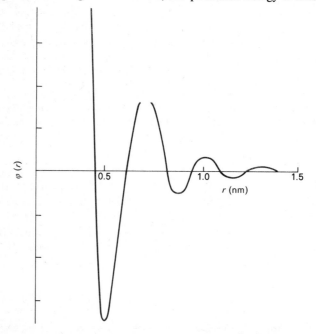

Fig. 14.2 The effective pair interaction function in liquid aluminium (after Chen, 1971).

of a direct couiombic interaction between ions (screened by electrons) and an indirect interaction between ions via electrons. Nevertheless, at least for the "simpler" metals such as Na, Al, Mg with small ion cores an effective pair interaction has been proposed; although unlike the potential energy functions of Chapter 11, in addition to

* Wilson, 1965, Adams *et al.*, 1967; March, 1968; Borland, 1968; Beer, 1972; Faber, 1972; Takeuchi, 1973.

the steep repulsion at short distances followed by an energy minimum, theoretical considerations predict damped oscillations at larger distances (Fig. 14.2). However, this modification of the effective pair potential energy function would have only a slight effect on density distribution functions, and experimental verification of the proposed oscillatory potential energy function is difficult to obtain.

Semiconducting liquids (xi)

Liquid semiconductors might be expected to fill a role intermediate between a liquid metal and an ionic liquid, and in fact examples occur in "impure liquids". Liquid vanadium pentoxide is an n-type semiconducting oxide in the temperature range 670–1000°C and at oxygen partial pressures greater than 0.1 atmospheres, as indicated by a positive temperature coefficient of conduction. Liquid sodium vanadates exhibit n-type semiconducting properties at sodium oxide concentrations less than 10 mol %, although ionic conducting properties are observed at sodium oxides concentrations greater than 25 mol %. It is considered that the conduction mechanism is delocalized electron ("polaron") movement between V^{4+} and V^{5+} centres for the semiconducting sodium vanadates, in contrast to sodium and oxide ion movement as charge carriers in the ionic liquid (Kerby and Wilson, 1972).

General references

*** ADAMS, P. D., DAVIES, H. A. and EPSTEIN, S. G. (1967), eds., *The Properties of Liquid Metals* (Proceedings of the First International Conference, Brookhaven National Laboratory, 1966), Taylor and Francis, London; also published as "Structure and Scattering in Liquid Metals", *Adv. Phys.* (1967), 16 (62, 63, 64), 147–307.

** BEER, S. Z. (1972), ed., *Liquid Metals Chemistry and Physics,* Marcel Dekker, New York.

* CLARKE, J. H. R. and HILLS, G. J. (1973), "Molten Salts – a Survey of Recent Developments", *Chem. Britain,* 9, 12–9.

COVINGTON, A. K. and DICKINSON, T. (1973), eds., *Physical Chemistry of Organic Solvents*, Plenum, N.Y.

*** DURRAN, T. H. (1971), *Solvents,* Chapman and Hall, London, 8th edn., revised by E. H. Davies.

FABER, T. E. (1972), *An Introduction to the Theory of Liquid Metals,* Cambridge University Press, London.

*** FRANKS, F. (1972), *Water: A Comprehensive Treatise,* Plenum, New York, Vol. 1, *The Physics and Physical Chemistry of Water.*

*** JANDER, G., SPANDAU, H. and ADDISON, C. C. (1967 on), eds., *Chemistry in Nonaqueous Ionizing Solvents,* Vieweg, Braunschweig, and Pergamon, Oxford, a series of volumes.

** LAGOWSKI, J. J. (1967, 1970), ed., *The Chemistry of Nonaqueous Solvents,* Vol. 2, *Acidic and Basic Solvents;* Vol. 3, *Inert, Aprotic and Acidic Solvents,* Academic Press, New York.

*** RIDDICK, J.-A. and BUNGER, W. B. (1970), "Organic Solvents: Physical Properties and Methods of Purification", in *Technique of Chemistry,* ed. Weissberger, A., Vol. II, *Organic Solvents,* 3rd edn., Wiley Interscience, New York.

*** TAKEUCHI, S. (1973), ed., *The Properties of Liquid Metals,* (Proceedings of the Second International Conference, Tokyo, 1972), Taylor and Francis, London.

*** TIMMERMANS, J. (1950, 1965), *Physico-Chemical Constants of Pure Organic Compounds,* Elsevier, Amsterdam, Vols. 1 and 2.

* WADDINGTON, T. C. (1965a), ed., *Non-Aqueous Solvent Systems,* Academic Press, London.

* WADDINGTON, T. C. (1969), *Non-Aqueous Solvents,* Nelson, London.

** WILSON, J. R. (1965), "The Structure of Liquid Metals and Alloys", *Metallurgical Rev.,* 10 (40), pp. 381–590.

Specific references

ADDISON, C. C. (1967), "Chemistry in Liquid Dinitrogen Tetroxide", in Jander *et al.* (1967), Vol. III, Part 1.

BLANDER, M. (1964), ed., *Molten Salt Chemistry,* Interscience, New York.

BLOOM, H. (1967), *The Chemistry of Molten Salts,* Benjamin, New York.

BOCKRIS, J. O'M. and REDDY, A. K. N. (1970), *Modern Electrochemistry,* Macdonald, London, Vol. 1, Chap. 6, "Ionic Liquids".

BORLAND, R. E. (1968), "Energy Levels of Electrons in a Liquid Metal", in *Physics of Simple Liquids,* eds. Temperley, H. N. V., Rowlinson, J. S. and Rushbrooke, G. S., North-Holland, Amsterdam, Chap. 16, pp. 693–710.

BUROW, D. F. (1970), "Liquid Sulfur Dioxide", in Lagowski (1970), Chap. 2, pp. 137–85.

CHEN, S.-H. (1971), "Structure of Liquids", in *Physical Chemistry: An Advanced Treatise,* eds. Eyring, H., Henderson, D. and Jost, W., Academic Press, New York, Vol. 8A, pp. 85–156.

DAVIS, M. M. (1970), "Brönsted Acid–Base Behaviour in 'Inert' Organic Solvents", in Lagowski (1970), Chap. 1, pp. 1–135.

DOVE, M. F. A. and CLIFFORD, A. F. (1971), "Inorganic Chemistry in Liquid Hydrogen Fluoride", in Jander *et al.,* Vol. 1, Part 2.

DRAGO, R. S. and PURCELL, K. F. (1965), "Co-ordinating Solvents", in Waddington (1965a), Chap. 5, pp. 211–51.

EISENBERG, D. and KAUZMANN, W. (1969), *The Structure and Properties of Water,* Oxford University Press, London.

FEHÉR, F. (1970), "Liquid Hydrogen Sulfide", in Lagowski (1970), Chap. 4, pp. 219–40.

GILLESPIE, R. J. and ROBINSON, E. A. (1965), "Sulphuric Acid" in Waddington (1965a), Chap. 4, pp. 117–210.

HORNE, R. A. (1970), "Water Properties", in *Kirk-Othmer Encyclopedia of Chemical Technology,* Interscience, New York, Vol. 21, pp. 668–88.

HORNE, R. A. (1972), ed., *Water and Aqueous Solutions, Structure, Thermodynamics and Transport Processes,* Wiley, New York.

HYMAN, H. H. and KATZ, J. J. (1965), "Liquid Hydrogen Flurode", in Waddington (1965a), Chap. 2, pp. 47–81.

JANZ, G. J. (1967), *Molten Salts Handbook,* Academic Press, New York.

JOHNSON, J. F. and PORTER, R. S. (1970), eds., *Liquid Crystals and Ordered Fluids,* Plenum, New York.

JOLLY, W. L. and HALLADA, C. J. (1965), "Liquid Ammonia", in Waddington (1965a), Chap. 1, pp. 1–45.

KELKER, H. (to be published), in series "Monographs in Modern Chemistry", *Liquid Crystals,* Verlag Chemie, Weinheim.

KERBY, R. C. and WILSON, J. R. (1972), "Electrical Conduction Properties of Liquid Vanadates. I. Vanadium Pentoxide. II. The Sodium Vanadates", *Canad. J. Chem.,* 50, 2865–70, 2871–76.

KILPATRICK, M. and JONES, J. G. (1967), "Anhydrous Hydrogen Fluoride as a Solvent and a Medium for Chemical Reactions", in Lagowski (1967), Chap. 2, pp. 43–98a.

KLANBERG, F. (1967), "Liquid Hydrogen Chloride, Hydrogen Bromide and Hydrogen Iodide", in Lagowski, (1967), Chap. 1, pp. 1–41.

LAGOWSKI, J. J. and MOCZYGEMBA, G. A. (1967), "Liquid Ammonia", in Lagowski (1967), Chap. 7, pp. 319–71.

LEE, W. H. (1967a), "Sulfuric Acid", in Lagowski (1967), Chap. 3, pp. 99–150.

LEE, W. H. (1967b), "Nitric Acid", in Lagowski (1967), Chap. 4, pp. 151–89.

LIND, J. E., ABDEL-REHIM, H. A. A. and RUDICH, S. W. (1966), "Structure of Organic Melts", J. Phys. Chem., 70, 3610–19.

LIND, J. E. (1973) in Advances in Molten Salt Chemistry, ed. Braunstein, J., Mamantov, G. and Smith, G. P., Plenum, New York, Vol. 2, pp. 1–26.

MAIER, S. and FRANCK, E. U. (1966), "The Density of Water Between 200 and 850°C and 1000 to 6000 bar", Ber. Bunsenges. Phys. Chem., 70, 639–45 (German).

MARCH, N. H. (1968), "Effective Ion–Ion Interaction in Liquid Metals", in Physics of Simple Liquids, eds. Temperley, H. N. V., Rowlinson, J. S. and Rushbrooke, G. S., North-Holland, Amsterdam, 15, pp. 645–91.

MOORE, W. J. (1972), Physical Chemistry, Longmans, London, 5th edn., Chap. 19.

PARKER, A. J. (1962), "The Effects of Solvation on the Properties of Anions in Dipolar Aprotic Solvents", Quart. Rev., 16, 163–87.

PAUL, R. C. and SANDHU, S. S. (1970), "Acyl Halides as Nonaqueous Solvents", in Lagowski (1970), Chap. 3, pp. 187–240.

PAYNE, D. S. (1965), "Halides and Oxyhalides of Group V Elements as Solvents", in Waddington (1965a), Chap. 8, pp. 301–52.

PEACH, M. E. and WADDINGTON, T. C. (1965), "The Higher Hydrogen Halides as Ionizing Solvents", in Waddington (1965a), Chap. 3, pp. 83–115.

POPOV, A. I. (1970a), "Anhydrous Acetic Acid as Nonaqueous Solvent", in Lagowski (1970), Chap. 5, pp. 241–337.

POPOV, A. I. (1970b), "Other Carboxylic Acids", in Lagowski (1970), Chap. 6, pp. 339–79.

RAHMAN, A. and STILLINGER, F. H. (1971), "Molecular Dynamics Study of Liquid Water", *J. Chem. Phys.,* 55, 3336–59.

REINSBOROUGH, V. C. (1968), "Physical Chemistry of Molten Organic Salts", *Rev. Pure and Appl. Chem.,* 18, 281–90.

REYNOLDS, W. L. (1970), "Dimethyl Sulfoxide in Inorganic Chemistry", in *Progress in Inorganic Chemistry,* ed. Lippard, S. J., Vol. 12, Interscience, pp. 1–99.

RHODES, E. (1972), "Fused Salts as Liquids", in *Water and Aqueous Solutions: Structure, Thermodynamics and Transport Processes,* ed. R. A. Horne, Wiley, New York, pp. 175–243.

RITCHIE, C. D. (1969), "Interactions in Dipolar Aprotic Solvents", in *Solute–Solvent Interactions,* eds. Coetzee, J. F. and Ritchie, C. D., Marcel Dekker, New York.

SHARPE, A. G. (1965), "The Halogens and Interhalogens as Solvents", in Waddington (1965a), Chap. 7, pp. 286–99.

SUNDHEIM, B. R. (1964), ed., *Fused Salts,* McGraw-Hill, New York.

SYMONS, M. C. R. (1972), "The Structure of Liquid Water", *Nature,* 239, 257–9.

UBBELOHDE, A. R. (1972), "Ionic Melts as Model Liquids", *Revue Roumaine de Chim.,* 17, 357–60 (English).

VAUGHN, J. W. (1967), "Amides", in Lagowski (1967), Chap. 5, pp. 191–264.

WADDINGTON, T. C. (1965b), "Liquid Sulphur Dioxide", in Waddington (1965a), Chap. 6, pp. 253–84.

WALRAFEN, G. E. (1968), "Structure of Water", in *Hydrogen-Bonded Solvent Systems,* eds. Covington, A. K. and Jones, P., Taylor and Francis, London.

WERES, O. and RICE, S. A. (1972), "A New Model for Liquid Water", *J. Amer. Chem. Soc.,* 94, 8983–9002.

Appendix:
Fourier transformation

This discussion (Doetsch, 1961; Bracewell, 1965; Champeney, 1972) will be introduced in terms of functions varying with time t or frequency ω, but can be applied also to any other variables, in particular momentum transfer $\hbar Q$ and position vector $\mathbf{r}$ in scattering experiments.

A *periodic* function $F(t)$ may be expanded in a *Fourier series*:

$$F(t) = a_o + 2 \sum_{n=1}^{\infty} (a_n \cos nt + b_n \sin nt) \tag{A.1}$$

where

$$a_n = \frac{1}{2\pi} \int_{-\pi}^{+\pi} F(t) \cos nt \, dt$$

and

$$b_n = \frac{1}{2\pi} \int_{-\pi}^{+\pi} F(t) \sin nt \, dt$$

for $n = 0, 1, 2, \ldots$. The Fourier series may also be written in the complex form

$$F(t) = \sum_{n=-\infty}^{+\infty} c_n \exp(int) \tag{A.2}$$

where

$$c_n = \frac{1}{2\pi} \int_{-\pi}^{+\pi} F(t) \exp(-int) \, dt \tag{A.3}$$

for $n = 0, \pm 1, \pm 2, \ldots$. It can be considered that $F(t)$ is formed by the summation over all values of n of the oscillations $\exp(int)$, with amplitude and phase expressed by c_n.

If on the other hand the event is *not periodic*, then the function $F(t)$ can be described by a *Fourier integral* instead of by a Fourier series:

$$F(t) = \int_{-\infty}^{+\infty} f(\omega) \exp(i\omega t) \, d\omega \tag{A.4}$$

where the Fourier transform function $f(\omega)$ sometimes denoted $\hat{F}(\omega)$, can be determined from $F(t)$ by the "minus-i" Fourier transform

$$f(\omega) = \frac{1}{2\pi} \int_{-\infty}^{+\infty} F(t) \exp(-i\omega t)\, dt \qquad (A.5)$$

The function $f(\omega)$ corresponds to the coefficients c_n, but instead of a sum over integral indices n, the integral over the continuous variable ω appears. Thus $F(t)$ is not built up from oscillations with integral frequencies, but oscillations of *all* frequencies are required, i.e. the variation with time (the spectral representation, $F(t)$) of a non-periodic event may be described by a *spectrum* covering all frequencies. The function $f(\omega)$ is called the density of the spectral distribution at the frequency ω, or the *spectral density*.

As an example, consider the time function

$$F(t) = \exp(-|t|) \qquad (A.6)$$

which is depicted in Fig. A.1a. The spectral density of this function is

$$f(\omega) = \frac{1}{2\pi} \int_{-\infty}^{+\infty} \exp(-i\omega t) \exp(-|t|)\, dt$$

$$= \frac{1}{\pi} \int_{0}^{\infty} \exp(-t) \cos \omega t\, dt$$

$$= \frac{1}{\pi(1+\omega^2)} \qquad (A.7)$$

Thus the Fourier transform of an exponential function is the Lorentz or Cauchy function, equation (A.7), shown in Fig. A.1b.

The nature of the spectral representation of the function $\exp(i\omega t)$ is very simple, a single spectral line or oscillation with frequency ω and amplitude 1, and with all other oscillations of zero amplitude, but the concept of the spectral density fails.

As is the case for ω and t, the static structure factor $S(Q)$ and the total correlation function $h(r) = g^{(2)}(r) - 1$ are related by the Fourier transformation

$$S(Q) = 1 + \rho \int [g^{(2)}(r) - 1] \exp(i\mathbf{Q} \cdot \mathbf{r})\, d\mathbf{r} \qquad (A.8)$$

In the same way that ω is the transform of t and t is the transform of ω, (equations (A.4) and (A.5)) there is reciprocity between $S(Q)$ and

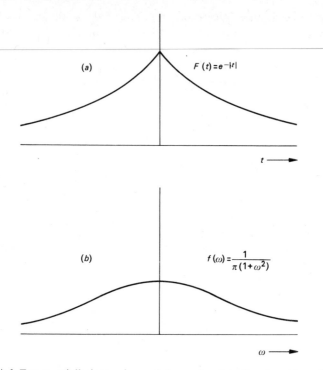

Fig. A.1 Exponentially increasing and decreasing time function (a), and the corresponding spectral density (b).

$g^{(2)}(r) - 1$: not only is $S(Q)$ the transform of $g^{(2)}(r) - 1$, but also vice versa.

The double Fourier transformation involving both t and ω, $\mathbf{r}$ and $\mathbf{Q}$, is

$$S(Q, \omega) = \frac{1}{2\pi} \iint \exp\left[i(\mathbf{Q} \cdot \mathbf{r} - \omega t)\right] G'(\mathbf{r}, t) \, d\mathbf{r} \, dt \qquad (A.9)$$

where $G'(\mathbf{r}, t)$ was introduced in equation (5.6) and $S(Q, \omega)$ is the dynamic structure factor. Similarly the self-dynamic structure factor $S_s(Q, \omega)$ is related to the self-correlation function $G_s(\mathbf{r}, t)$ (equation (5.7)) by

$$S_s(Q, \omega) = \frac{1}{2\pi} \iint \exp\left[i(\mathbf{Q} \cdot \mathbf{r} - \omega t)\right] G_s(\mathbf{r}, t) \, d\mathbf{r} \, dt \qquad (A.10)$$

The spectral density $\hat{z}(\omega)$ of the velocity autocorrelation function $z(t)$ is

$$\hat{z}(\omega) = \frac{1}{2\pi} \int \exp\left(-i\omega t\right) z(t)\, \mathrm{d}t \qquad (A.11)$$

The infra-red absorption band shape $I(\omega)$ and the dipole moment autocorrelation function are also related by Fourier transformation:

$$I(\omega) = \frac{1}{2\pi} \int \exp\left(-i\omega t\right)\langle \mathbf{p}(0) \cdot \mathbf{p}(t)\rangle\, \mathrm{d}t \qquad (A.12)$$

$$\langle \mathbf{p}(0) \cdot \mathbf{p}(t)\rangle = \int \exp\left(i\omega t\right) I(\omega)\, \mathrm{d}\omega \qquad (A.13)$$

In nuclear magnetic resonance spectroscopy, the spectral density of the correlation function of the local magnetic field $\mathbf{H}$ produced by molecular motion is

$$I(\omega) = \int\limits_{-\infty}^{+\infty} \langle \mathbf{H}(0) \cdot \mathbf{H}(t)\rangle \exp\left(-i\omega t\right)\, \mathrm{d}t \qquad (A.14)$$

General references

BRACEWELL, R. (1965), *The Fourier Transform and Its Applications,* McGraw-Hill, New York.

CHAMPENEY, D. C. (1972), *Fourier Transforms and their Physical Applications,* Academic Press, London.

DOETSCH, G. (1961), *Guide to the Applications of Laplace Transforms,* Van Nostrand, London.

Notation guide

Symbol	Meaning	Equation or Chapter where first used
a	van der Waals attraction parameter	2.6
a	constant in Stokes' Law	3.12
a	(subscript) attractive	2.17
A, A'	empirical parameters or functions	
Å	(unit abbreviation) angström, 10^{-10} m	
b	van der Waals repulsion parameter	2.6
B, B'	empirical parameters or functions	
B_j	virial coefficient	2.4
c	concentration	3.2
c	speed of light, $2.997\,925 \times 10^8$ m s^{-1}	
c	(subscript) coulombic	11.1
c	(subscript) critical	4.1
cm	(unit abbreviation) centimetre	
C	(unit abbreviation) coulomb	
$c(r)$	direct correlation function	5.11
d	(subscript) distinct	5.7
d	(subscript) dispersion	11.3
D	diffusion coefficient	3.2
D_o, D'_o	parameters in empirical equations for D	3.14
D	(unit abbreviation) Debye $(3.335\,64 \times 10^{-30}$ C m$)$	8.6
$\mathbf{D}$	electric displacement	8.1
$-e$	electronic charge, $1.602\,19 \times 10^{-19}$ C	3.7, 8.7
eV	(unit abbreviation) electron-volt, $1.602\,2 \times 10^{-19}$ J; $(Le\mathrm{V} = 96,487$ J mol$^{-1})$	
E	parameter in Arrhenius equation	3.14
$E, \mathbf{E}$	electric field strength	3.6, 8.1
δE	energy transfer	6.3
F	Faraday constant, $9.648\,67 \times 10^4$ C mol^{-1}	3.6
$\mathbf{F}_i$	fluctuating intermolecular force on molecule i	Chapter 12

g	parameter in square well function	11.9
$g^{(2)}, g^{(2)}(r)$	pair distribution function	5.2
$G(\mathbf{r}, t)$	van Hove distribution function	5.5
$G'(\mathbf{r}, t)$	van Hove correlation term	5.6
h	Planck constant, $6.626\,20 \times 10^{-34}$ J s	
$\hbar$	$h/2\pi$	
H	molar enthalpy	2.10
$h(r)$	total correlation function	5.10
i	(subscript) induced	11.3
i	(subscript) typical molecule	
I	electric current	3.6
$I(\omega)$	electromagnetic absorption as a function of frequency	Chapter 9
j	(subscript) typical molecule	
j	(subscript) number of virial coefficient	2.4
J	flux	3.2
J	(unit abbreviation) joule	
k	Boltzmann constant, $1.380\,62 \times 10^{-23}$ J K^{-1}	
k	scattered wavenumber	6.2
k_o	incident wavenumber	6.2
kbar	(unit abbreviation) kilobar (10^8 Pa)	
$\mathbf{k}$	scattered wave vector	6.1
$\mathbf{k}_0$	incident wave vector	6.1
l_p	dipole length	8.7
L	Avogadro constant, $6.022\,17 \times 10^{23}$ mol^{-1}	
m	effective molecular mass	7.7
m	parameter in pair potential energy functions	Chapter 11
m	(unit abbreviation) metre	
M	molecular weight	3.13
n	refractive index	8.4
n	parameter in pair potential energy functions	Chapter 11
n	running co-ordination number	Chapter 5
$n^{(1)}$	singlet distribution function	5.1
$n^{(2)}, n^{(2)}(r)$	doublet distribution function	5.2
$n^{(3)}$	triplet distribution function	Chapter 5
nm	(unit abbreviation) nanometre, 10^{-9} m	
N	number of molecules	
N	(unit abbreviation) newton	

p	pressure	2.1
$\mathbf{p}, p$	dipole moment	Chapter 6
pm	(unit abbreviation) picometre, 10^{-12} m	
Pa	(unit abbreviation) pascal, N m^{-2}	
$\mathbf{P}, P$	shear stress	3.1
$\mathbf{P}$	electric polarization	8.2
$\mathbf{Q}, Q$	wave vector transfer	6.1
Q_o	value of Q at S(Q) peak	Chapter 7
r	radius	3.12
$r, \mathbf{r}$	intermolecular distance	5.1
r_o	intermolecular distance of minimum energy	11.12
r	(subscript) reduced	4.1
R	gas content, 8.314 3 J K^{-1} mol^{-1}	
s	(subscript) self	5.7
s	(subscript) electrostatic	11.2
s	(unit abbreviation) second	
S	spin angular momentum quantum number	Chapter 9
S	molar entropy	7.10
$S(Q)$	static structure factor	7.1
$S(Q, \omega)$	dynamic structure factor	7.3
S	(unit abbreviation) siemens ($=$ ohm^{-1})	
t	time	
T	(absolute) temperature	2.1
T_0	lower temperature limit of mobility	3.15
T_1	n.m.r. longitudinal relaxation time	Chapter 9
T_2	n.m.r. transverse relaxation time	Chapter 9
u	mobility	3.4, 3.7
U	molar internal energy	2.9
ΔU_{vap}	molar energy of vaporization	2.19
$\mathbf{v}, v$	velocity	
v	velocity of sound	7.8
V	molar volume	2.1
V_f	molar free volume	Chapter 3
z	ionic charge number	3.7
$z(t)$	velocity time autocorrelation function	5.9
$\hat{z}(\omega)$	spectral density of velocity autocorrelation	7.4
α	molecular polarizability	8.5
α, β	empirical parameters	
β	dimensionless energy variable, $\hbar\omega/kT$	Chapter 7

β	isochoric thermal pressure coefficient	2.12
γ	friction constant	7.7
δ	solubility parameter	2.20
ε	molecular energy parameter	11.9
ε	permittivity	8.1
ε_0	permittivity of a vacuum	8.2
ε_r	relative permittivity	8.3
ζ_i	friction coefficient	3.11
η_s	shear viscosity	3.1
κ	electrical conductivity	3.6
κ_T	isothermal compressibility	7.1
λ	wavelength	
Λ	molar conductivity	3.8
Λ_0, Λ_0'	parameters in empirical equation for Λ	3.14
μ	chemical potential	3.3
ξ_i	molecular friction coefficient	Chapter 12
π	internal pressure	2.12
ρ	number density	5.1
ρ_1, ρ_2	intermolecular distance parameters	11.15
σ	scattering cross-section	Chapter 7
σ	molecular distance parameter	11.6
τ	relaxation time	
τ	time spent in molecular free diffusive movement	7.7
τ	time interval in Kirkwood theory	Chapter 12
ϕ	fluidity	3.14
ϕ_0, ϕ_0'	parameters in empirical equations for ϕ	3.14
$\phi, \phi(r)$	molecular potential energy functions	11.1
Φ	potential energy of molecular configuration	12.2
ω	frequency	
ω_0	frequency of incident radiation	7.9

Name index and subject index

Name Index

Entries in italics are general references not specifically cited in the text. The abbreviation (ed.) indicates "editor".

153

Subject Index